FOREWORD BY AUSTRALIAN FEDERAL
SENATOR MALCOLM ROBERTS

CITIZEN ONE

I0092411

THE
CASE AGAINST
DIGITAL IDENTITY

PAUL G CONLON

BEng (Hons), BIT

Citizen One: The Case Against Digital Identity
© Paul G Conlon 2024

ISBN: 978-1-923197-17-6 (paperback)
        978-1-923197-18-3 (eBook)

A catalogue record for this book is available from the National Library of Australia

www.paulgconlon.com
www.oceanreevepublishing.com
Published by Paul G Conlon and Ocean Reeve Publishing

REEVE
PUBLISHING

About the Author

Queensland-based author, Paul G. Conlon, is a professional engineer and product manager with a unique perspective on the intersection of identity and technology. His formative years were shaped by his family's survival of Nazi Germany's identification policies, igniting an enduring passion for understanding the historical impact of centralised identity systems.

With degrees in Electronic Engineering and Information Technology, Paul has two decades of global experience working with digital systems in the private sector and is a former Engineering Officer in the Australian Army. In 2023, he was selected as a judge for 'hack:DiD', a global software competition and long-term initiative dedicated to empowering self-sovereign identity (SSI) education. In 2024, he was invited to write a speech on the dangers of digital ID, to be delivered in the Australian Senate.

Paul's exploration of centralised identity's historical consequences resonates deeply in today's digital realm, where government-controlled digital ID poses new threats to individual privacy and freedom. Paul's ability to explain complex systems in simple terms, honed through commercial-technical writing and daily news articles, covering financial and political trends, makes him a powerful voice in the debate over the future of digital ID.

Foreword

One Digital ID to rule them all? One book to add to the top of your reading list.

Are you becoming aware of increasing surveillance and control from government and big tech? For years, I've been warning about a single national government-run Digital Identity scheme. Now that it's become a reality, Australians need to know how it will impact their lives and what's at stake.

As a fellow engineer, I recognise that Paul Conlon has made a good case against the Federated Digital Identity scheme which the Australian Government legislated in May 2024. The case is even stronger when you learn that this so-called Trusted Digital ID framework has been left intentionally weak. The protections on data storage and use are almost non-existent. Feedback during the consultation phase which pointed out privacy and security failings built into the infrastructure were ignored.

It is perilous to have been thrust, as a nation, into a situation where the Government decides the purpose for our data and who gets to have access to it, only after its collection. Trust should be grounded in transparency and accountability, rather than be presumed. Inherent integrity is not a term that goes hand in hand with government, whether Liberal or Labor.

Top-down government mandates to impose policies such as a national Digital Identity scheme rather than bottom-up grassroots initiatives indicate the control mechanism behind them. We saw this during the COVID response. Did medical tyranny lay the groundwork for digital tyranny?

The author's own family history allows him to draw on Nazi Germany as an historical reference. This book methodically lays out the well-trodden path that has led us to this point, covers what increased centralised power over identity management will likely mean for our future, and offers practical advice for digital minimalism to ensure your privacy, security, and freedom.

Australians have a choice about the kind of world we want for our children. Think of Citizen One as a self-help guide for navigating the digital future.

Forewarned is forearmed.

—Australian Federal Senator Malcolm Roberts

Dedication

Dedicated to Beulah, Layla, and Edison, for giving me
a reason to write.

Contents

Acknowledgements

To my dear wife Beulah, as these pages come to life, I am deeply grateful for your unwavering support and profound love. Your grace and strength, along with those whose suffering this book chronicles, have been my guiding lights, and it is with heartfelt appreciation that I dedicate this work to you.

During challenging times, such as when drafts were lost to the digital abyss or fatigue set in, your patience and understanding tempered my responses and steadied my course. Your invaluable wisdom, imparted as you attentively listened to the endless drafts and ideas that surfaced at all hours, has profoundly shaped both this book and its author.

You navigated the Herculean complexities of our daily lives with exceptional dedication, providing me with the resources to pursue this quest for human betterment. Your sacrifices did not, and never do go unnoticed. To the contrary, they are the silent pillars supporting this endeavour and all that will follow. This book stands as a testament to your incredible support and a tribute to the amazing journey we are on together. Thank you for being the most extraordinary partner in life and love.

To Loretta Davis, hammering my drafts out on the anvil of your literary experience gave me insights into refined writing that may have otherwise taken years to acquire. Your honesty and directness are rare and align well with the purpose of this book. The critiques you shared are sure to aid readers in understanding the vital content on the following pages and enrich their lives. Thank you, Loretta, for your generosity. I owe a tremendous debt of gratitude.

To Wayne Jean, if the world had more men like you—champions of liberty and staunch rejectors of tyranny—we'd find ourselves navigating far fewer problems. Your support, evidenced by your insightful suggestions, provision of reference books, and regular encouragement, has been invaluable. Receiving your written feedback, invariably drenched in red ink, was always a highlight, transforming the daunting into the doable. Thank you, Wayne, for being the embodiment of the change we wish to see in the world.

To Wim Hochepied, your companionship in the creative trenches was what I needed, when I needed it. You have taught me to find hope amidst the darkness and infuse these pages with an optimism that stands defiant against the malice they describe. I am grateful for your candour, in alerting me when my work fell short, and in acknowledging when it shone.

I'm also thankful for the contributions of Aman Jabbi, Keisha Gamman, and the many unseen warriors—from politicians to artists—currently battling against the predation of government-issued human identification systems. Here's to the difference our united efforts will make.

To you, the reader. It is for you that these words have been woven together. Crafted to inform, inspire, and spark change, your engagement and curiosity give this book its purpose. Together, we champion a future that upholds freedom and integrity in the digital age. Thank you for being a vital part of this mission.

Lastly, I would also like to honour those who, in the eternal battle against government identity harvesting, have lost lives and livelihoods alike. I humbly hope this work brings meaning to your suffering.

Prologue

"Danke ... bitte ... guten Tag." [Thanks, please, good day].

Four simple German words ... that irrevocably altered the course of history for my family. Four simple German words, hastily thrown together by a US Army soldier during his mission to liberate Berlin in 1945. His rifle was pointed directly at a terrified young German woman, his bloodied finger quivering over the trigger. That young woman was to become my beautiful and gentle grandmother, my *Oma*.

The soldier struggled with the dilemma before him. He struggled to *identify* her. Was she just a German, who should be slain like the many others who had stared down his barrel over the prior weeks? Or was she simply a young woman caught in the crosshairs of a horrifying war, deserving of a chance at life? Did her identity span both possibilities?

Identity is a matter of life and death in government hands. To the Nazi government, otherwise known as the *Third Reich*, my *Oma's* fate was of no concern. She was just a number, stamped on a mandatory ID card years prior, and left to the whims of invading soldiers. Are we forgetting history's lessons? How do they apply to you in the digital ID age?

Today, we are sleepwalking into a societal restructure based on uneven information sharing between ourselves and our governments. A structure where our identities are becoming transparent to government, while government is becoming opaque to us. A structure that we now, like Germans then, would not have chosen had we been paying more attention.

This book exposes the enduring justifications for centralised identity registration and their predatory outcomes. It challenges the belief that "this time it's different." Together, we will explore the essence of identity, trust, technology, and governmental power, revealing the simple fact that some will always seek to exploit these for harm.

But amidst these horrors, this is really a story about love over coercion. My family history shows how love can fulfil our collective responsibility to resist the datafication and commodification of our identities, and it is my honour to share it with you ...

CHAPTER 1:
Lessons from my Oma

Ensuring we do not move into a world where digital identities replicate existing bias and prejudice in the non-digital world should be a critical threshold issue for the whole Parliament.

—Australian Federal Senator David Shoebridge, Senate Economics
Legislation Committee Dissenting Report, commenting on
the Australian Digital ID Bill 2024. (1 p. 80)

If the Government sends me an identification card, I shall return it with a letter to explain that I do not need it;

I already know who I am.

—Douglas Graham (Tasmania), Australian newspaper contributor protesting
the 'Australia Card', 8 September 1987. (2 p. 261)

Don't ... trust ... governments.

—My paternal grandmother (Oma). Born 1922 (Germany)-Died 2012 (Australia).

My Family: Born in Conflict

Let me tell you a love story. It's a story born from the conflict of the Second World War. It's the story of my family, and it starts with my beloved paternal grandmother, my *Oma*. Born in Germany in 1922, she was a woman of substantial strength, traditional morals, and unwavering principles. She shared with me her life experiences, and fostered in me a passion for individual liberty, responsibility, and, above all, self-identity.

Her father, Fritz, my great-grandfather, was a veteran of the First World War. He had been decorated with the prestigious Iron Cross for heroism at the 1917 Battle of Passchendaele in Belgium. Under artillery fire, he had bravely climbed a tree to sight enemy positions. Having lost parts of his abdomen doing so, he was no stranger to suffering (Figure 1).

By the Second World War, Fritz had risen to the rank of major in the *Wehrmacht* [the armed forces of the German Third Reich] (Figure 2). Now a father, he lived with my great-grandmother, Charlotte, whom he'd married in 1920, along with my *Oma* and her younger brother, Manfred. The family home was in a German town called *Ratibor*, at the time, near the Polish border.

The War's Start

I vividly remember my *Oma* sharing stories of those days. When I was young, she relived the events of September 1st, 1939, when her country's mechanised vehicles of war unexpectedly rumbled through her town, towards Poland. As a teenage girl, the smell of exhaust fumes and the rattle of the *Panzer* [tank] tracks left an indelible impression on her. She had experienced the genesis of the Second World War firsthand.

Nine months later and a mere 1,100 kilometres to the west, a separate struggle was unfolding. There, on the beaches of Dunkirk, stood my

Figure 1. My great grandparents, Fritz and Charlotte, on their wedding day, 1920. Fritz's Iron Cross is visible.

Figure 2. My great grandparents, twenty years later, in Nazi Germany, 1940. Fritz's Iron Cross is still visible.

paternal grandfather, Bill. He found himself caught in the relentless onslaught of the Nazi *Blitzkrieg* [lightning war].[1] Serving as a warrant officer in the British Royal Signals Corp, he had been deployed to France with the British Expeditionary Force, in response to Germany's invasion in 1940.

My grandfather had yet to meet my *Oma*. Indeed, he had no idea that fate would lead him to love and a fulfilling post-war life. Although destined to witness the growth of his grandchildren into adulthood, he and his brothers-in-arms lived every minute as their last. The gates of hell had swung wide open, and the seemingly unstoppable Nazi war machine had barged straight through. Like my *Oma*, nine months earlier, the sights, sounds, and smells of that experience left an indelible impression on him.

---

[1] Lightning war was an effective military tactic used by the *Wehrmacht* to quickly overwhelm opposition with targeted use of tanks, infantry, artillery, and air support.

Communication played a critical role in these engagements, with signals units often being the first to arrive and the last to leave the battlefield. Hence, my grandfather became one of the last of over 300,000 British and French soldiers to be evacuated from the beaches of Dunkirk between May 26 and June 4 in 1940. (3 p. 529) While others were rescued, he remained highly exposed on the flat beach, relentlessly strafed by the *Luftwaffe* [German air force] as he performed his duties. To this day, his bravery remains a source of great pride, and serves as a personal beacon of defiance against evil that I strive to emulate.

My grandfather was certainly not a man of violence. Bill exuded warmth and gentleness, and like my *Oma*, was a victim of circumstance. He rode motorcycles and horses in impressive displays of balance and riding talent, hosted by the Royal Signals. These events which followed the transition from horseback to motorcycles, were an opportunity to showcase *esprit de corps* (Figure 3).[2]

Figure 3. *My paternal grandfather, Bill, pictured towards the rear with fellow British servicemen circa 1930s, Royal Signals motorcycle display.*

---

[2]   A sense of adoration, comradery, and dedication to a group with a common cause. Common in military units.

The War's End

Fast forward to 1945—the war in Europe was drawing to a bloody conclusion. My German relatives were living in Berlin, where my great-grandfather, Fritz, had been stationed. By this time, my *Oma's* brother, Manfred, had fallen in Operation Barbarossa, the war against the Soviet Union (Figure 4). My family had tragically lost its youngest member, the country was in ruins, and countless thousands of people were homeless and hungry.

Figure 4. *My Oma's younger brother, my grand-uncle, Manfred. 1924-1944. Killed on the Russian front, aged just twenty.*

As the Allied Forces advanced on Berlin, Nazi resistance had weakened to the point where children, identified via the government's population registry, were being conscripted for frontline defensive duties. In a deplorable disregard for humanity, German civilians were seen as mere expendable resources by their government. In one instance, my *Oma* had been instructed to shoulder a bazooka—as she described it—in defence of the city.[3] She was unable to even lift it.

The expectation that young women would shoot at tanks in city streets highlights the vast reach of Nazi Germany's propaganda efforts. This colossal disinformation machine, masterminded by Joseph Göbbels—who we'll meet later— not only terrorised individuals, but, more disturbingly, also manipulated the collective psyche of a nation. In the years leading up to this point, the government had worked tirelessly to instil fear throughout the German population via many tactics—a key one being the portrayal of the approaching enemy as savages.

African-American soldiers were particularly demonised in this way. The government's fabricated tales of horror had been sufficient in both grandeur and frequency, to firmly entrench the fear of these so-called animals into the German mindset.

In a remarkable twist of fate, and arguably the catalyst for this book, my then 22-year-old *Oma* came face to face with these fears one day amidst the chaos as Berlin fell. With her father occupied by his military duties, my *Oma* had been left with her mother, Charlotte, to fend for themselves. The women found refuge from the war-torn streets by sheltering together in a basement as Allied troops took Berlin.

She recalled how the Nazi government's propaganda had raced through her mind. Like a scene from the movie, *Downfall*, the two vulnerable women huddled in that dark basement, protected only by shadows and prayer, awaiting their fate as the vicious sounds and smells of war inched ever closer outside.[4]

---

[3]   Unfortunately, details of the specific weapon were not divulged before my *Oma's* passing. It was likely she could not identify it at the time. From available information, it was likely the 88 mm *Raketenpanzerbüchse 54* anti-tank rocket launcher commonly known as *Panzerschreck* [tank fright].

[4]   *Downfall* (German: *Der Untergang*), 2004, directed by Oliver Hirschbiegel.

Suddenly, the basement door burst open with a thunderous kick, showering the room with flying splinters of wood. Startled, the women looked up. Blocking the doorway at the top of the basement staircase stood a backlit figure, shrouded in smoke. To their horror, they realised that this was the dishevelled personification of what they had been indoctrinated for so many years to fear most—an African-American man wearing the hated US Army uniform.

My *Oma's* mouth went dry, as wide-eyed, she stared directly into the eyes of her Grim Reaper. The weight of her imminent death bore heavily upon her fragile shoulders. Her heart quivered with a cocktail of complete terror and devastating sorrow. Her mind was instantly engulfed with thoughts of unfulfilled dreams, stolen hopes, and the piercing ache of farewells unsaid. *Manfred!* she thought, *I'll see you soon.*

It is surreal to me now that my very ability to write this book came down to the split-second decision of that soldier at that precise moment. For an instant, the fate of my entire family line rested in his blood-stained hands. Scanning the room to find only the two women hiding there, the soldier's battle-hardened face softened as thankfully, he arrived at a moral choice. Deciding to label the women victims, not perpetrators, he mustered the only German words he knew to soothe their fear before moving off:

Danke, bitte, guten Tag. [Thanks, please, good day].

Little did he realise that those four simple words spoken in that basement on that day would far outlive the terrified women who heard them. They did their job in communicating that his intent was not to slaughter women, yet they also conveyed a far more fundamental lesson. Those four words hit my *Oma* with a metaphorical force, every bit as powerful as the physical ordnance exploding outside. It was the force of reality clashing against the lies of her government.

In a stellar example of how the truth can set you free, one man's hobbled-together parting gesture had shattered the cumulative propaganda of one of the strongest military empires the world had ever seen. Absolutely everything that the Nazi government had inflicted on their people up to that moment, including compulsory identity registration, had been sold as a necessary protection against evils that had just proven themselves benign. Again, in a very visceral sense, my *Oma* experienced the end of the Second World War in Europe, just as she had its beginning.

A Dance of Reconciliation

The end of the war did not mean the end of suffering. Across Europe, daily existence remained arduous and uncertain. Amidst the settling dust and the fading echoes of battle, the German people grappled with the harsh realities of rebuilding their shattered lives. My family found their place in the intricate fabric of this challenging era, weaving their own modest contribution.

Fritz survived the war and had been reunited with the family. My *Oma*, and her parents, were living under British occupation in *Vienenburg* as they tried to process the loss of Manfred, their beloved brother and son.[5] Due to the shortages, an arrangement was in place where families would share accommodation, and food was distributed through a rationing system.

My *Oma's* family cohabitated with a man by the name of Knieriem who happened to be a skilled *Dolmetscher* [interpreter].[6] By virtue of his profession, Knieriem had an association with, and worked for, the British Commandant in charge of the occupied area where they lived. My *Oma* described the commandant to me as 'progressive', a man who understood that the current living situation was unsustainable in the longer term.

---

[5]   *Vienenburg*, a borough of Goslar, capital of the Goslar district, Germany.

[6]   At the time, this vocation was considered distinct from (and more prestigious than) a translator.

Several factors contributed to this, including the imbalance of supplies and the generally held 'us and them' mentality that segregated the people, and hindered post-war integration.[7, 8] Consequently, the commandant solicited suggestions on ways to address these issues, and ease the feeling of occupation held by the German people in his region. Knieriem suggested that a good way to begin would be to hold a social ball and invite the locals, along with members of the occupying military. The gathering represented a symbolic olive branch that was funded and catered by the British.

The commandant initially expressed hesitation at this suggestion, doubting any expectation of German attendance. In response, Knieriem noted that he shared accommodation with a German family—my family—that was well regarded in the community.[9] He offered that, if they went, then he would be confident members of other German families would also attend, ensuring the event's success.

Thus, the ball received the green light. The commandant, keen on maintaining decorum, personally oversaw the evening's proceedings. My *Oma* and her mother, pivotal in garnering public enthusiasm for the event, were even honoured with seats at the distinguished head table alongside him.[10]

One of the soldiers in attendance at that first ball was my grandfather, Bill. My *Oma* recounted that he, nine years her senior, with an unmistakable air of British refinement, graceful, and impeccably presented, was the only one who approached, asking her to

---

[7] The military seemed to have better access to food, for example.

[8] The presence of foreign troops understandably created a sense of friction and discontent within the local community.

[9] The family had been affluent before the war, owning a vineyard that provided employment. They had also often engaged in community philanthropy which earned a good reputation for my great-grandfather in particular, who was a known Iron Cross recipient.

[10] My *Oma* could not recall the British Commandant's name as she told me this story from her death-bed in palliative care. If fate happens to place this book in the hands of his descendants, I would be most interested in making contact.

dance. She likely appeared off-bounds given the seating proximate to his commanding officer, so this venture by a sole warrant officer would have required both confidence and initiative. After the success of the ball where my grandparents met and shared their first dance, my family was instrumental in facilitating other similar events in the community.

The ball symbolised a glimmer of hope and new beginnings, yet it was set against the backdrop of a nation grappling with its identity. The joy my grandparents found in each other's company was in stark contrast to the widespread despair and hardship that the German people faced as they rebuilt their lives from the ruins.

The post-war era was an extended period of mental and physical difficulty, requiring effort from those such as Knieriem and his insightful commandant. Indeed, my *Oma* confessed that the ball had provided an opportunity to augment the family rations—by discreetly stuffing food into her pockets. The suffering and fear of the German people during the war, and the years that preceded and followed it, have not been an area of historical focus.

History tends to spotlight the justifiably gallant actions of liberating armies and their supporting cultures. However, it perhaps poorly features the suffering and gallantry of those forgotten citizens within the borders of runaway authoritarian regimes. In the case of Nazi Germany, ordinary Germans had to contend with enemies both foreign and domestic. They were equally at war with their own government, as they were with troops abroad.

The resilience demonstrated by my grandparents in the face of tyranny is not just family lore; it represents the universal struggle for control over our destinies. These personal narratives powerfully testify to the broader themes we will explore, illustrating how individual lives are inextricably linked to the grand currents of how we express ourselves in the world.

Central Themes

There are three significant themes that dance in my mind when I think of the stories my grandparents shared with me about those days. These are the themes we will explore, at length, in this book.

Theme One: The Tyranny of Human Commodification
Identification and Surveillance as Weapons

The central theme revolves around humanity's perennial struggle against identity-based tyranny. In my *Oma's* youth, nationalised identity unveiled who people were, and what they could do, while nationalised surveillance laid bare where they were, and what they were doing. Their merger, enforced by the police state, severely curtailed freedom, and eroded the very fabric of life.

My grandparents wished to spare their descendants from the suffering they had seen this reality produce; a legacy they shared with countless others.[11] They saw their struggles as dire warnings for future generations, alerting them to the horrors that ensue when governments develop a taste for commodifying the people they supposedly serve. This foresight grants the youth an opportunity, although not a guarantee, to sidestep the visceral suffering that occurs on both sides of a government's border, when the equilibrium of knowledge between it and its population, tips in favour of those in power.

This book's purpose is to delve into that disequilibrium, and specifically, to reveal the predation of centralised government identity registration, warts, and all. It uncovers how the Nazis shrouded the construction of just such a system with ornamental justifications,

---

[11]   We will visit this noble characteristic again when discussing how deployed Australian volunteers were against attempts to introduce Australian conscription during the First World War. This was on the grounds that they knew what suffering the conscripts would be involuntarily subjected to.

later weaponising it to inflict unimaginable horror on its own people, and the world at large.

Despite entering from opposite sides of that war, my grandparents emerged with a consistent analysis. The profound truth they learned, borne from personal experience, was clear. In my own words:

> *With absolute knowledge of a population's identity*
> *comes absolute command of its destiny.*

The Apparent Protections of Modern Technology

As we delve into the historical parallels of digital ID systems, one might briefly pause to wonder whether there are contemporary examples that mirror these concerns? How do modern digital ID systems differ from their historical counterparts, and what safeguards are in place today?

While the historical misuse of identity registries by authoritarian regimes raises valid concerns, proponents of modern digital ID systems argue that today's technology, enhanced with democratic oversight and stringent laws, differs significantly from the past. They emphasise the potential benefits such as enhanced security features, reduced bureaucracy, and increased convenience, contending that these systems merely restructure existing government data, posing no new risks.

Advocates believe that robust safeguards and accountability mechanisms can effectively protect citizens' rights, highlighting the importance of considering both the risks and the potential advantages of these systems. While this book aims to explore these opposing views, it must be done within the historical context of how data has been weaponised when checks and balances fail. The chilling efficiency of Nazi Germany in using identity data for oppressive purposes serves as a relevant backdrop.

The Historical Context

These are arguments that have been used to dismiss similar concerns before. Yet, the catastrophic consequences of past data misuse justify why concerns about government digital ID cannot be dismissed lightly. Nazi prowess in domestic and foreign identity harvesting, tracking, and statistical analysis remains impressive today.

As early as 1936, Friedrich Burgdörfer, as the Director of the Office of Statistics, used population census data to aid the German military.[12] His calculations concluded that the regime would be well-placed to wage war against Poland and France combined by 1940. These figures predicted that by that time, more than 300,000 Germans of adequate utility would reach military draft age each year. (4 p. 28)

Why should census data be of any concern? This seemingly benign information gifted the *Wehrmacht* with strategic foreknowledge, which they exploited to precisely time their invasion of Poland, resulting in a devastating loss of young lives. Burgdörfer's analysis, which was off by a mere four months, reveals the ominous power conferred by absolute knowledge.

The Nazis meticulously tracked individuals on index cards, creating a level of insight unimaginable to most.[13] Tools like the *Volkskartei* [German population registry], proposed to Hitler in 1934 and later forced on the population, allowed the government to know more about its people, than the people knew about their government. This pernicious asymmetry produced constant fear, stemming from the victims' total blindness to the information held about them by the government.

---

[12] Born April 24th, 1890, Burgdörfer was instrumental in furthering German statistical analysis. He notably established the 1925 'economic and social-statistical evaluation' census. It was the first to capture data on an individual's physical and mental fragility (4 p. 16).

[13] A system proposed by Freiburg Lawyer Erwin Cuntz which we will explore in Chapter Four.

Collecting identity data was so crucial to Nazi political aims, that census takers closely followed the *Wehrmacht* into newly occupied territories to identify and segregate people. Statisticians stole identity from the living, just as soldiers had stolen breath from the dead. Once the system had been established, there was simply no escaping it.

A Timeless Struggle for Freedom

But this is no academic study. My *Oma* matured in an era when power dictated justice; when behaviour was sanctioned by badges on uniforms, and pistols on belts; when so-called security police broke girls' arms over nuances in their parents' identity cards, laughing while doing so. It was a time when identifying as Aryan, was the currency by which survival was temporarily purchased from officialdom.

It is often said, "You don't know what you've got until it's gone" ... and there is no more painful loss than that of individual freedom. My *Oma's* moral obligation was to pass these lessons of liberty to me, just as it is my moral obligation to pass them to you. As my *Oma* would often say (and she was not alone):

Protect your privacy and your freedom.
They are your greatest treasures.

Her words echo a chorus of historical figures, who have also felt the sting of tyranny. Take Patrick Henry, one of America's Founding Fathers and the first governor of Virgina, who urged us to:

Guard with jealous attention the public liberty,
and suspect everyone who comes near that precious jewel.

Henry lived through an era when proving loyalty to the British Monarch, rather than genetic purity to a dictator, was the currency by which survival was temporarily purchased from officialdom.

Raised in 1980s Australia, I, like millions of others like me, have inherited the precious jewel of freedom, earned by the toil and sacrifice of gallant men and women who bled before us. Yet, does inheriting the unearned pose a risk of undervaluation? Intergenerational lessons are as fragile as snowflakes in the midday sun; we tend to dismiss them, only to relearn them anew after they have melted.

Theme Two: Eternal Vigilance

Being Alert

The second theme closely mirrors the first and focuses on the enduring cost of liberty: eternal vigilance. The defeat of Hitler's particular brand of tyranny did nothing to secure a struggle-free future. On the contrary, persistent efforts to abuse humanity through identity registration are nearly inevitable. Examples have continually surfaced throughout history, and many remain active today. Each attempt comes cloaked in appealing slogans and serves to exploit the social anxieties of the day.

Despite these disguises, the unmistakable hallmarks of human stocktaking remain constant: appeals to efficiency and virtue, manipulation, de-humanisation, laws, and punishments. Recognising these recurrent patterns serves to guard against apathy, which blinds us to the slow but steady encroachment of such systems.

Tyranny is a patient foe. When it resurfaces, its oxygen is charitable-sounding legislation, and its food is propaganda. It will patiently wait decades to creep up on you in the shadows while you sleep. It claims to be the alibi of every generation's collective safety, and yet views us with contempt and our children with disdain. Regardless, it needs our assistance, relying on incremental public policy, and ignorant compliance, to have palpable effect.

So pervasive are the ongoing attempts to centrally manage identity, that my grandparents faced the threat twice in their lives. Having narrowly escaped Europe's identity scourge, they came face to face with the same predator over four decades later, on the other side of the world. The 'Australia Card', proposed in the 1980s, demonstrated how a country of alert and invested individuals bridged political, economic, and social divisions to protect yet another generation from a similar fate to Germany.

Our Modern Obligations

Today it is our turn. On our watch, governments continue to extend their influence into ever more aspects of our lives. Global institutions are standardising their policies, and increasingly using force in their approach, all while trust in democratic processes continues to fall.

The contemporary tyrant's toolkit is slick and modern. Technological advancements have opened the door to a level of totalitarian control that is now more invasive than ever. In these dangerous circumstances, no facet of social life can thrive—be it in economics, culture, science, technology, or arts, the state-sponsored centralisation of human identity, is progressively paralysing an otherwise prosperous time.

Are we truly ready for digital ID? Its adoption amidst the emerging Artificial Intelligence (AI) and Machine Learning (ML) phenomena remains unproven. Concerns are growing over its proposed use to police the internet against increasingly eclectic threats. In our society, driven by convenience and hedonism, especially in an era dominated by social media and smart devices, we risk navigating these uncharted waters blindly.

The Historical Context

Yet we are not completely without guidance. The warnings from 1930s Germany are pertinent today for those with eyes to see.

This was not an ancient culture from which we have fundamentally evolved. The Second World War transpired less than a century ago—a mere two or three generations. This was not a culture that had lost its moral compass; Germany was a profoundly Christian society at the time. Nazi registration and census policies were not a historical aberration. Rather, they parallel the digital ID and surveillance protocols governments are hurriedly adopting today, seemingly without care, caution, or concern, let alone debate, discretion, or deliberation.

An individual's *competencies*, once known by the Nazis, condemned them to mandatory work assignments considered best aligned to the war effort. In a very real sense, they became slaves, unable to control the most basic parts of life, such as where to live, and when to wake up for work. Nazi statisticians produced benchmarks for expected human work output. If an individual's productivity fell short of that anticipated, using metrics based on a deep knowledge of their identity, they were considered deficient. Malnourished slaves, German or otherwise, deemed to be working too slowly in Nazi factories were "beaten to death." (5 p. 14) Today, we offer these observations willingly, with cloud connected Fitbits and live security camera feeds from within our homes and workplaces.

In contrast, an individual's *deficiencies*, once known by the Nazis, condemned them to sterilisation or murder.[14] Nazi population statistics coldly identified those people of net economic value to the government, and those who were not. This metric, in lieu of any appreciation for individual qualities, was all that mattered once identity registration had sufficiently de-humanised the population. Today, many are compelled to advertise deficiencies and disabilities to the government, in return for payments or perks.

---

[14] Estimates of those sterilised under the *Law for the Prevention of Genetically Diseased Offspring* range from 250,000 to 500,000 (15 p. 61).

Theme Three: Love Offers a Solution

Moving Forward Together

The third, and perhaps most personal theme from the story of my grandparents' union, is that love will prevail, but only if we allow it (Figure 5). My *Oma* personally witnessed the February 1945 bombing of Dresden.[15] She jumped from a train on the outskirts of the city and ran for cover under trees, to watch in horror while 3,900 tons of explosives incinerated, asphyxiated, and buried alive up to 250,000 men, women, and children.[16, 17, 18] The resulting 1,600 acres of firestorms alone are thought to have killed tens of thousands of people (6).[19]

Despite this, she married and deeply loved an honourable man who once fought under the same Union Jack worn by the Lancaster and Mosquito bomber pilots, high above her that night. With love, great things are possible. Whether they be bombs falling from the sky or worse, digital ID, great evils can be overcome not through apathy, but by finding the good in humanity, spreading awareness, and building alternative outcomes and solutions.

Love enables us to set aside our petty differences and focus on building a future worth living for, one where the need to delegate the determination of trust to the government is obsolete (Figure 6). With love, we can get to know our communities personally, rather than via government decree; which brings us to my thesis.

---

[15] The horror of this experience is not easily articulated. One attempt however comes from Eleonore Kompisch who was in Dresden during the bombings. *"Until the bombing of Dresden I was very religious, but after that I lost my faith completely. I could not believe any more."* (92)

[16] Charred bodies, hardly recognisable as human, littered streets. Some were atomised, like petrol in a carburettor, beyond physical form.

[17] The fierce building fires sucked oxygen from cellars below, asphyxiating people sheltering there.

[18] This estimate accounts for the influx of undocumented refugees from the Eastern front (6).

[19] An area equivalent to 6.5 square kilometres.

Figure 5. My paternal grandparents, circa 1950s. A relationship born from the war.

Figure 6. My paternal grandparents, circa 1990s. After the defeat of "The Australia Card".

Thesis

This book's central argument is that we are unwittingly heading towards a cultural transformation characterised by an uneven exchange of information between citizens and their governments. In this emerging framework, our personal information is increasingly accessible to the state, while the workings of government are being increasingly hidden from us.

Likewise, we are ceding control to major corporations that mandate identification for their products and services. These entities can shape our behaviour by restricting access to vital resources, such as social media platforms that serve as modern public squares for free speech, simply for seeing the world differently. The hardships resulting from the vaccine mandates of recent years exemplify how such power can impact those with differing views.

Time to Wake Up

Modern society has fallen victim to a collective failure of imagination. Together, we have been lulled into indifference, easily distracted from the true direction that governments and corporations are seemingly herding us. We have failed to, or perhaps we refuse to, imagine what may happen if we pursue our current path. Equally, we simply don't see what could be if we choose a better one. Instead, rational debate is moving further outside the Overton Window, the range of politically acceptable discussion.

The literature of our time has an obvious gap. We need a broadly palatable body of research to appeal to the middle ground if we are to gift our children the same freedoms gifted us by my grandparents' generation. We need to lovingly rouse those asleep to the very real dangers we face in the coming months and years. This requires knowledge, for knowledge is power. You are now holding that body of knowledge in your hands.

This thesis asserts that historical mass identity registries have led to catastrophic outcomes, and today's digital versions, which we will explore in Chapter Four, could potentially do so again. Exploring this thesis successfully requires consideration of how current generations interact with technology.

Millions of people today undervalue anonymity, casually submitting their DNA to companies that claim to offer "real science, real data and genetic insights." (7), (8) They do this, seemingly unaware of the grave historical consequences of revealing genetic data, clearly shown by the repercussions of the *1933 Law for the Prevention of Genetically Diseased Offspring.* (4 p. 102)

The loss of privacy, especially genetic privacy, through these types of registries and biometric databases, has resulted in real suffering, with people losing their lives, children facing unspeakable atrocities, and entire lineages being wiped out. It is our fundamental duty to critically examine the infrastructure and laws being implemented right now, rather than giving them a mere cursory glance.

Many find comfort in the belief that laws safeguard rights in the digital age. However, Dr John Lennox, a scholar and author at the crossroads of science, philosophy, and religion, cautions that the reliability of these safeguards is uncertain without guidance from a universal moral standard:

> *Hitler, in his political youth, made treaties*
> *but he tore them up once he got into power. (9)*[20]

Even in democratic societies, there is potential for governments to misuse data, particularly during crises when extraordinary measures seem justified. Laws are simply formalisations of power, mirroring the perspectives of those able to inflict them. The resulting temptation for control can undermine the very democratic principles designed to protect citizens.

Furthermore, treating others by consensus may be somewhat functional in cultures built on equal centres of power. Each can withhold value from the other to ensure negotiation occurs. Yet it is easily argued that the modern world has enabled great power imbalances. In such a case, one would expect similar asymmetry to be paralleled in the laws passed in such cultures.

A law can equally forbid, or mandate murder, but it cannot explain why either is preferable. The best approach is therefore to prevent the construction of technical systems that are dangerous when legislation fails. We are comfortable making this argument with AI, yet we largely fail to do so where digital ID is concerned. *The choice to not build infrastructure that may be weaponised by government is itself also an act of democracy.* The collective inaction of thousands

---

[20]   Spotify interview with Dr Jordan Peterson titled *A Conversation About God*. The quoted section starts at 1:06:00 into the discussion.

of engineers and technicians represents a choice to not allow these powerful tools to be placed in the hands of the already powerful.

What Do People Really Want?

Some government digital ID advocates often claim strong public demand and support, but evidence suggests otherwise. Research by Redfield & Wilton Strategies in July 2023 indicates that the economy, taxation, immigration, crime, and election integrity are primary concerns for voters. Digital ID did not even make the list, which is surprising given that *"don't know"* was included as a category. Moreover, eighteen percent of respondents identified government spending as a concern, hinting that the cost of implementing and running digital ID could exacerbate rather than ease voter concerns. (10)

A related argument could be made that digital ID can reduce environmental impact by minimising the resources associated with physical documents, thus reducing the footprint of identity management. However, it is important to weigh this against the energy consumption and environmental toll of supporting large-scale, centralised digital infrastructures like always-on data centres and networks.

If we value the environmental impact of our systems and institutions, then we are best served by addressing the low-hanging fruit first. We must focus on the savings realised when the government abdicates its resource hungry identity ambitions. The cost of living and housing pressures represent the most visible of many contemporary struggles, making the need to be sparing with funds increasingly important. A more effective approach to reducing environmental impact then, might be to significantly cut government bureaucracy. This tactic could potentially reduce significantly more waste than introducing new bureaucratic measures, thinking that aligns far better with Redfield & Wilton's voter feedback.

Anecdotal observations support these findings. Everyday conversations frequently revolve around the cost of living and immigration, yet there is a notable absence of discourse on the need for digital ID. Not once have I heard someone raise the issue as their life's main priority. This prompts the important question:

Does the legislative push for digital ID truly reflect the public's organic will, or is it a policy driven by officials, and passively accepted by a population preoccupied with more immediate concerns?

Such questions are particularly important to ask when historical attempts to impose centralised government identity management, such as the Australia Card in the 1980s, were proven to use manufactured public support as a manipulative tool.[21]

Even if we concede a high demand for digital ID exists, it underscores the need for free market competition to foster a competitive environment with superior products. Government solutions tend to be slow, costly, and inferior compared to free-market offerings, where providers jostle for voluntary customer patronage by offering the best identity products. Unlike mandated solutions, which don't need to prove their efficacy due to a lack of alternatives, free market options must demonstrate their value to attract and retain users.

Our Journey

The What

Within the following pages, we will embark on a journey. Its path exposes the essence of identity, reveals the nature of trust and technology, and converges on the true face of government, revealing a litany of sins that must not be forgotten. This exploration is interwoven with the fabric of my family's lived experiences, offering a

---

[21] We will justify the use of identity registration as a manipulative tool in Chapter Six.

poignant lens to see how identity has become a canvas upon which governments wield a formidable brush.

Together, we will delve into the real-world manifestation of identity and the significant effects of its exposure, including resource denial and vulnerability to exploitation. This exploration will deepen our self-knowledge and shed light on government efforts to eliminate our anonymity. We challenge the idea that governments are the sole architects of societal trust and order. Instead, we expose the dangers of centralised judgments of character, highlighting the misuse of labels, such as 'terrorists' and 'undesirables' for political ends.

We will learn that trust, essential for society, should not be delegated to an influential few, swayed by donors and intoxicated with power. Trust should instead grow organically within the community it benefits.

Using original type-written Nazi documents, the perils of government-led identity initiatives that claim to foster trust become clear. These ideas often come disguised as utopian visions of safety, security, and convenience. Yet the interpretation of our identity data required for these unobtainable dreams can shift with political climates, leading to retrospective judgments of innocence.

Amidst these sombre reflections, we will also uncover rays of positivity. Entrepreneurs and engineers who, in the vast expanse of the identity space, are crafting alternatives to this centralised future. In what's known as self-sovereign identity (SSI), we find a beacon of hope, a life raft for those who dare to champion a future where individuals can reclaim and retain control over their own identities online.

But I must caution that this book will most certainly change you. By the end, you will see the world with a fresh clarity, and appreciate that dark forces have always seen our identity as the key to impose their political and social will at scale. Despite the Australian flavour, this journey may be superimposed upon any Western nation and beyond, all of which currently push similar agendas, albeit under differently named legislation.

By the conclusion, and with your help, digital ID's predation will become a cornerstone of everyday discussion. You will become a custodian of important truths that come with a commensurate obligation to share. It may also come with difficulties, as most pursuits of value do.

The How

As we delve into these narratives, it is important to reflect on the tone we adopt. The urgency and intensity that permeate our discourse are not without purpose. In times when society's vigilance wanes and the stakes are high, a forceful articulation is not merely a stylistic choice, but a moral imperative. Such a tone may draw criticism, often labelled as alarmist or conspiratorial, yet this is a small price to pay if it adequately conveys the gravity of the situation.

If history has taught us anything, it is that complacency can lead to catastrophe. Therefore, if our warnings are heeded and we err on the side of caution, that is a victory. But if our concerns are vindicated, then the strong tone we have chosen will have been not only necessary but perhaps even lifesaving.

Accusations of alarmism are common techniques against those who challenge mainstream narratives, a topic we will crystalise when discussing labels in Chapter Three. Labelling dissenting or unpalatable views in this way is an *Ad Hominem* attack, an emotional reaction when deeply held beliefs are challenged. This linguistic tactic criticises the person, or their motives, instead of engaging with the argument itself, revealing a lack of rational thought. As such, these labels can generally be dismissed as something other than logical, civil discussion.

Furthermore, in an era dominated by 'fake news' and 'fact-checking', such terms serve to abruptly halt discussion based on emotion, not dispassionate facts. They are the rhetorical equivalent of pulling the fire alarm at a graduation dinner. One would hardly remain in one's chair, continuing to eat, until evidence of the alarm's validity materialises. The cry, "Fake news!" sees people rush to the exits; they ask questions only later, if at all. This approach does not resolve debate but highlights the fact that more is needed.

Such tactics can also be attributed to a misunderstanding of time horizons. Months and years may pass without any manifestation of the warnings in this book. Yet we do not know where we fit in the broader cycle of human behaviour over decades and centuries. The accusation of alarmism fails to recognise this book as a risk mitigation for possibly unlikely yet severe outcomes. It fails to acknowledge that, given sufficient time, encountering these events is likely inevitable.

Advocating for seat belts with graphic warnings might seem alarmist to those who have never experienced accidents. Yet the risk of encountering a drunk driver looms at every intersection. Normalcy bias, the idea that what happened yesterday will continue tomorrow, does nothing to protect us from the next identity abusing regime, perhaps lurking just around the corner.

The Why

The German philosopher Friedrich Nietzsche once wrote:

He who has a why to live can bear almost any how.

So why should the rejection of centralised, government controlled digital ID be of interest to us? What makes it worth sacrificing an evening with our favourite streaming services, or a weekend fishing with friends to consider? It is because, as we shall explore, these systems threaten the very freedoms that permit us to stream video and go fishing with friends.

Knowledge of identity is possession of the means to define what that identity is, and what it is permitted to do. Without the freedom to keep our identities to ourselves, all other topics of discussion are academic. By handing possession of our identities to a centralised digital authority, rather than retaining them for ourselves, we render all other pursuits meaningless. If you value your streaming and fishing, you by default value the possession of your individual identity.

What are our choices?

We are poised at the precipice of a decision that will change civilisation. The first option is to embrace and firmly claim title to the countless qualities that comprise our individual identities, and consequently the sovereignty, to decide our own futures. The second option is to collectively abandon our identities to a monolithic, centralised digital database, where our every thought, feeling, and action is surveilled, and 'one size fits all' policies and punishments are imposed without hope of redress.

Liberty's flame is fragile. We need to make this decision quickly, as freedom is not passed to our children in the bloodstream. It is like an Olympic torch, held and protected by every generation's run, before being handed to the next, for them to do the same.

My intention with this book is to keep that vulnerable flame of liberty alive; to keep my *Oma's* cautionary tales where they belong, in the past. We all have a moral obligation to point out threats to that precious jewel.

My hope is that you arrive at the same conclusions as a particular British warrant officer, and the daughter of a German major once did. Our tolerance for the injustice of government-imposed digital ID should be paper thin. If we truly live in a society that values inclusion and equality, as we constantly claim, then now is the time to put talk into action.

For these reasons we should be interested in digital ID. Its hurried and unchartered rollout is a Trojan Horse that risks extinguishing liberty's flame for generations, if not forever. Drawing from the lessons of history, we can envision a scenario—as speculative as it may be—where, left unchecked, government identification and surveillance provide the tools for tyrants, present or future, to deprive our children of life's essence.

This is not a prediction but a distinct possibility, and one that has previously become a reality. Behind the bars of future gulags, should such a grim future come to pass, your enslaved children will wonder why their parents' love for them was insufficient to fight against such policies, when there was still time.

This book may help spark enough good souls into action to avoid that future. It may just save our children's lives. There are moments in history that call upon people of the lesser nobility to perform feats of the greater purpose. I extend an invitation to join me on this quest, and should you accept, breathe deeply, and turn the page. Things are about to get dark ...

CHAPTER 2:
Identity

*Every act of extermination was preceded by an act of registration;
selection on paper ended with selection on the ramps.*

—Götz Aly and Karl Heinz Roth, The Nazi Census: Identification and
Control in the Third Reich. Written 1983. (4 p. 1)

*"Nothing to hide, nothing to fear," directed at each member of the
public, should be turned around and directed at government as:
"No legitimate reason to know, no legitimate reason to ask."*

—Paul Chadwick, Victorian Privacy Commissioner,
The Value of Privacy, 23 May 2006. (11)

*Nothing can be more un-Australian than the need to prove one's identity
on the call of an official, be it a policeman or a bureaucrat ...
... I refuse to be numbered and branded.*

—Australian Labor Party MP, Mr Lewis Kent, speaking in the House of
Representatives about the proposed Australia Card, 1986. (12 p. 181)

*The free and lawful person recognised by the common law
attains legal personality simply by attaining adulthood, and is
able to do anything which is not proscribed by law.*

*The de facto effect of compulsorily requiring adult citizens to
register with the state, is to introduce an additional element
which is essential for their legal personality.*

—Geoffrey Walker, Professor of Law and Head of Department,
University of Queensland, contributing to the Joint Select Committee
on the Australia Card 1986. (12 p. 132)

Nadine's Story

Digital ID is simultaneously a captivating social experiment, and profoundly one of the most dangerous. In this chapter, we explore the complexities of human identity, its importance, and the risks of centralised entities controlling its registration. Yet to understand these implications, we must first understand Nadine Schatz. The role that identity played in her life, and subsequently in her tragic death, is an instructive story best started with the significance of her name.

Schatz, meaning 'treasure' (in German), symbolised the value Nadine's Jewish parents undoubtedly placed upon her. Born in Paris just eight years after my *Oma*, Nadine embodied potential and promise. Her dark hair, mischievous smile, and inquisitive eyes painted a vivid picture. With fluency in two languages, her linguistic skill formed part of a unique identity; one that would never mature beyond childhood (Figure 7).

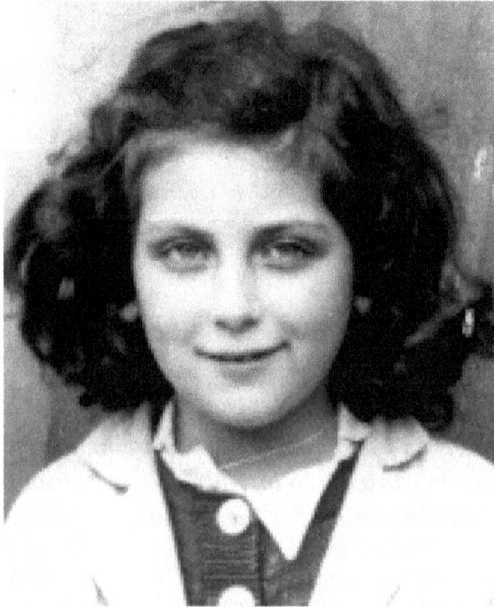

Figure 7. Nadine Schatz. Murdered thirteen days after her 12th birthday in an Auschwitz gas chamber, shortly after her arrival there on September 23, 1942. (13)

Just two days after the invasion of Poland, Nadine's home country, France, declared war on Germany. Then Prime Minister Edouard Daladier delivered an emotional appeal to the nation, resounding with determination and honour: [22]

> *In rising against the most frightful of tyrannies,*
> *in honouring our word, we fight to defend our soil,*
> *our homes, our liberties…*
>
> *… I greet with emotion and affection our young soldiers,*
> *who now go forth to perform the sacred task which we*
> *ourselves did perform before them. (14)*

Despite these impassioned words, Nadine's family grew sceptical of their government's resolve and sought refuge by moving to a coastal town near the Loire River. There, they lived peacefully until May 1940, when invading Nazi troops reached their town.

Yet France was a major collaborationist state, with some officials disgracefully aiding their occupiers by enforcing anti-Jewish laws. Rather than rising against the tyranny that Daladier had warned of, French police instead arrested Nadine and her mother in 1942, forever separating them. Nadine was deported through a Parisian transit camp to *Auschwitz*, arriving on September 23. (13)

Auschwitz

Auschwitz, Hitler's most notorious extermination camp, and the first to be corporatised, was a hellish place unfit for any child. Initially a concentration camp established near Krakow in June 1940, it later expanded into three sub-camps. Skilled workers arriving in boxcars, were forced into slave labour at *Auschwitz III*, run by chemical conglomerate *IG Farben*.[23] (5 p. 22) There they were forced to build and

---

[22] Edouard Daladier was ultimately imprisoned in Germany from 1943 until the end of the war. (3 p. 131)

[23] The conglomerate operating Auschwitz included Bayer. This global enterprise is still active today. (15 p. 62)

run a large manufacturing complex making synthetic rubber, aircraft fuel, munitions, and poisonous gases, all made profitable by the war. (15 p. 62)

Auschwitz I was the main accommodation camp. Its wretched living conditions offered only fleeting respite for the walking corpses by night. Moving between these sub-camps, people faced an average life expectancy of a mere three months. (5 p. 22)

Lacking any usable skills, however, young Nadine was deemed unfit for labour by *IG Farben* selection staff upon her arrival. She was transferred to *Auschwitz II, Birkenau*, which housed the gas chambers and crematoria. (5 p. 22) There, mass human extermination had commenced four months earlier. (3 p. 534) Shockingly, employees of a tax-exempt private company under the Nazi regime had sentenced her to death, with a mere pen stroke and hand wave. (16 pp. 81-82)

With military efficiency, Nadine was forced naked into a gas chamber alongside infants, the elderly, disabled, pregnant and sick. (3 p. 38) These strangers were to be companions through the last horrific experiences of her short life. Sealed inside, their collective fate was irrevocably determined. As *Zyklon B* gas filled the space, it extinguished all that could have been.

Within minutes, Nadine fell to the floor with nausea and headaches, followed by seizures and vomiting. (17) The ringing in her ears harmonised with the surrounding screams. Her little heart stopped three billion beats before its time, having never experienced love's first flutter.[24] Her eyes, those windows to the soul, clouded with the milky haze of death, never to see her own offspring.

---

[24] This estimate is based on an average heart rate of 80 beats per minute over a life span equivalent to my *Oma's*.

Nadine's Crime

Nadine's vibrant spirit and untapped potential were cruelly extinguished by humanity's darkest nature. Her body was discarded, tossed in an oven, and reduced to ashes that blanketed the countryside. Her crime? Simply being born to parents whose identities had been registered with the government, and in some small way, labelled as undesirable.

Being Jewish, among other things, earned someone this fate by the standards of the time. Today, it could equally have been supporters of certain outspoken political candidates, or people from specific regions currently in military conflict. Nadine's family religion was an identifying characteristic that had been 'leaked' into the world. She had lost sovereignty of the many personal traits comprising her identity, and they were used against her. Her life was not defined by her parents' faith, but as an identifying characteristic, that faith had the power to end it.

The Ongoing Global Tragedy

Nadine's story illustrates the devastating and irreversible consequences of being marked as 'undesirable' by the government. Yet the perils of merging state power with identity management echo not just in history, they persist today in our digital age, where personal information can be weaponised even more effectively. This was evident in September 2023, when my in-laws were performing missionary work in Sri Lanka. There, Christian refugees fleeing church burnings and physical attacks in Pakistan were facing hostility and deportation.[25]

Why, I pondered, *would refugees disclose their Christianity in such a hostile environment?* The answer proved straightforward. For many, their religion was documented in their passports and

---

[25] The United Nations High Commissioner for Refugees (UNHCR) was said to be concluding its operations in Sri Lanka, at the time.

official papers. In seeking asylum, faith had provided a lifeline, yet ironically, once officially registered, this same characteristic had become a curse.

Beyond mere statistics, my in-laws recounted the plight of a specific Christian family they met. With a young child in tow, they were confined in squalor. The wife, intimidated by male guards, was separated from her husband, their interactions limited to fleeting glances through the detention centre fencing. Like many others, the rejection of their asylum request was crushing, and forced a return to the dangers they had fled.

These stories, and millions more throughout time, showcase the profound and often negative impact of identifying data in the hands of the powerful. They echo my grandparents' warnings about what is possible once a nation's citizens are catalogued. Most importantly, they highlight the inextricable link between our identities and our physical existence.

Human Life: A Philosophical Discussion

Understanding digital identity begins with grasping the essence of identity itself. Reflecting on the moment I stood over my *Oma's* open coffin, I was struck by the contrast between her once vibrant spirit, and her now lifeless form. This moment underscored a universal truth: we are born with a unique essence that fades at death. Just as the sun traces a temporary arc across the sky, our lives are transient journeys that leave only memories and mementos, mere echoes of a human identity.

Exploring Humanity's Essence

Understanding human identity has long captivated philosophers, theologians, and scholars. This quest, championed by figures like Socrates, is seen in cultural markers like the Greek aphorism on the Temple of Apollo: 'Know thyself'. [26] Opinions vary, with René Descartes famously stating, in 1637: "I think therefore I am." Regardless, identity remains an enigmatic part of life.

Many see identity as an intrinsic and timeless essence, distinct from our physical form. Medical accounts of near-death experiences support this view. Intrinsic health advocate, Dr Zach Bush, describes cases where patients who die, with no brain, heart, or respiratory function, sometimes inexplicably return to life with their core identity intact. (18) These are known as near-death experiences. Notably, individuals retain memories, language skills, and political views, so prompting the question: where did those characteristics go after death? Retention of identity beyond physical form challenges the idea of reducing our complex selves to a single digital ID, which may fail to capture the intangible qualities that endure beyond what is seen.

Conversely, entrepreneur Alex Hormozi, who specialises in operationalising behavioural patterns for commercial success, suggests that identity is simply "internal culture." (19) This perspective likens identity to a set of predictable behavioural rules within an individual, much like the behavioural rules that define the culture within an organisation. Such thinking aligns with the experience of getting to know someone through predicting their reactions from their traits and past behaviours.

We acknowledge that our identities shape our conduct, and are influenced by our surroundings, yet we are not mere automatons responding to stimuli. Like poker players and politicians who expertly conceal their intentions, humans inherently wield unpredictability. This capacity to defy conditioned behaviours underscores the complexity of categorising people based solely on identity traits, especially in light of free will. Viewing identity as an 'internal

---

[26]   Originally, the Greek aphorism was *'gnōthi seauton'* (from the Temple of Apollo at Delphi).

culture' suggests that digital ID could enforce a uniform cultural mould on individuals, erasing the personal nuances that inform our demeanour.

These diverse views lead to conflict over our place in the world, and how we express ourselves in it. Discrepancies can risk exploitation and abuse as we navigate the complexities of defining identity, its ownership, and the social structures and policies that influence it. Clearly, humans are complex, and the term 'identity' is an oversimplification that fails to capture the full spectrum of our expression.

This single word not only bears the weight of describing us in the physical world, but is also now the sole measure by which governments assess our societal fit. Because identity is as varied as grains of sand, the term is too imprecise for anything beyond casual discourse. Despite this, we have based our society on this single concept, thereby overlooking the personal uniqueness that drives our decisions and actions.

Appreciating these nuances was my *Oma's* parting gift to me—she considered her identity to be her most valued possession. Her experiences in Nazi Germany made her guarded about revealing it to others, having seen firsthand how government registration was used to define, control, and terminate life. Moreover, surviving a time when life was considered cheap, allowed my *Oma* to appreciate the transcendence of each human existence. She spoke of our collective duty to view ourselves as more than numbers on computer screens and statistics in government publications. It falls on us to cherish and safeguard our identity, whatever that is, from the demons that relentlessly nibble away at it through actions both tame and bold.

Exceptional Qualities Demand Exceptional Treatment

Labelling ourselves as transcendent acknowledges our unique human qualities. As a species, we explore boundaries beyond our physical existence by seeking spiritual enlightenment and purpose beyond mere survival. Our ability to proactively assess potential actions against ideal standards sets us apart, allowing us to consciously shape our lives for better or worse.

Examples can be seen everywhere. When a tree falls on a house during a storm, for instance, we view it as a natural disaster without malice. The homeowner wouldn't blame the tree or seek compensation from nature. However, if an amateur arborist carelessly fells the tree, causing damage, they *could* be held liable and face resentment. This contrast underscores our accountability, showing how our thoughtful actions, and inactions, carry consequences.

Our exceptional qualities demand exceptional consideration. Even within our own species, we recognise that different standards apply due to cognitive and other limitations. The legal system accommodates this by considering factors like mental illness and impairment in its judgments. Yet, being human isn't only about facing harsher penalties; it's also about granting ourselves unique freedoms and rights in recognition of our enlightenment, and susceptibility to suffering.

Falling trees may evade judgment, but they're also incapable of experiencing life as profoundly as we do. Unable to love or pursue self-improvement, they meander through life like leaves in the wind, and are never mourned at open coffins. A tree is merely a collection of characteristics that does not suffer the impacts of objectification as deeply as we do. Its fate is determined by age, size, species, condition, and location. Reducing a person to such traits is different, serving only to overlook their potential and endanger their humanity.

Human Commodification: A Betrayal of Divine Rights

This is no abstract sentiment. The ramifications of identifying humans with loosely defined characteristics reverberate throughout history. A stark example comes from a 1941 German study by Professors Heinrich Kranz and Siegfried Koller. It proposed solving the perceived threat from anti-socials in Nazi Germany by:

"... taking away their cultural and national civil rights." (4 p. 99)

In Nazi Germany, being labelled as anti-social led to egregious human rights abuses such as sterilisation, forced abortion, child confiscation, marriage annulment, slave labour, and even murder. Conditions like epilepsy, mental disabilities, or behaviours like alcoholism and depression, all understandable at the time, were enough to warrant these punishments. Furthermore, the scholarly belief that these traits were hereditary, meant that relatives of the accused also faced similar fates.

The scope of this systemic identity-driven abuse warrants attention. Within twenty-two months of Hitler redefining euthanasia as the termination of 'lives without value' on November 1, 1939, more than 70,000 victims of mental illness were murdered. (3 p. 148) Towards the war's end, human sterilisations conducted by complicit doctors stabilised at around 65,000 per year. (4 p. 104) In today's terms, this equates to nearly 1.2 percent of Sydney's current population.[27]

These granular decisions were based on a deep and intimate understanding of family dynamics, extending even to the languages parents spoke at home with their children. This level of detail was crucial for the Nazi regime's racial policies, which aimed to categorise and segregate people based on their perceived racial purity and intrinsic social worth.

A revealing insight into this process comes from a secret classified report prepared by the Racial Political Office in Berlin, dated November 25, 1939. The report meticulously outlines the criteria

---

[27] This is based on the estimated residential population according to 2023 Australian Bureau of Statistics (ABS) data. (97)

for evaluating individuals and families, emphasising the importance of German heritage and language as markers of loyalty and racial purity. It stated:

Cases also come into consideration here in which people of German blood did not let their children learn German and spoke Polish in the family despite their German descent.

People in these intermediate classes who appear to be of low character are in any case to be considered as persons to be deported, regardless of whether they appear to be Germanisable or not.

This also includes those classes who are described as incapable of socialising, as anti-social, or whose low ability to survive indicates an accumulation of poor performance talents. (20)

A pervasive fear of such government determinations marked the era. These were shaped by legal instruments like the *1933 Law for the Prevention of Genetically Diseased Offspring* (4 p. 102), the *1934 Law for the Simplification of the Health System* (4 p. 104), and the *1935 Marriage Health Law* (4 p. 103). Such seemingly benevolent legislation, enacted years prior, morphed into tools of domestic terror. They legalised government intrusion into life based on arbitrary characteristics, and a view that those who held them were in some way diseased. [28]

Citizens lived every day looking over their shoulders. Castration was a very real consequence of the government knowing too much about you. While reflecting on the historical consequences of reducing humans to identity traits, it is important to also question how modern digital ID could likewise commodify people. *Could digital ID turn unique human lives into datasets ripe for exploitation and control?*

---

[28] The names of these legislative instruments offer important lessons. Positive sounding terms such as 'simplification', 'health', and 'marriage', hide the true nature of sterilisation and murder. A similarly critical eye should be cast upon the names of all legislative instruments. One cannot, after all, judge a book by its cover.

The Cost of Safety

Horror stories aside, it is crucial to acknowledge that psychopaths are not exclusively found within governments. Members of the public will also pose a risk. This fact gives birth to the perspective that government oversight, via digital ID, could play a role in enhancing security.

This counterargument posits that government control of identity data is the path to crime prevention, relying on surveillance and data monitoring for protection. Yet, this approach begs the question of the cost of such security. When unknown bureaucrats demand we trade personal freedoms now, for potential safety tomorrow, we must consider this a devil's bargain that invites a range of new risks. These include the potential for a disastrous breach in centralised government systems, leading to extensive identity theft and fraud, ironically making the 'cure' worse than the disease.

Identity centralisation compromises integrity by sharing all there is to know with another entity. This makes it nearly impossible to distinguish between the true owner and an identity thief—in the event of a data breach—because the attacker holds the same amount of data as the legitimate owner. There are also concerns about the self-policing nature of asymmetrical surveillance, a scenario echoing Benjamin Franklin's warning:

> *Those who would give up essential Liberty,*
> *to purchase a little temporary Safety,*
> *deserve neither Liberty nor Safety.*

In contrast, decentralised systems, those without a central point of control which we'll explore in the following chapter, provide strong security measures while avoiding the risks associated with centralised power, and intrusive monitoring. Like identity, security is a nuanced issue, concerning not only the definition of protection but also its degree, who provides it, and their intentions. Julian Assange, an Australian editor, publisher, and activist, once defined security as a spectrum, not an absolute. Striving for total security may compromise the very freedoms we are trying to secure, reminding us not to throw the baby out with the bathwater.

Treasuring Your Identity

Identity meant far more to my *Oma* than simple self-preservation. To her, personal questions needed to be justified by personal connections, and asking them for reasons of bureaucracy or curiosity was viscerally intrusive. Knowledge of her was a privilege earned through time, mutual respect, and trust. Like fitness or reputation, it could not be forced.

True security lies in maintaining control over one's identity. The instinct to safeguard this knowledge sees some turning to digital ID as a protective measure. Advocates argue that, with adequate security, it could more effectively mitigate selected types of identity fraud than vulnerable physical documents can. Although digital ID can potentially reduce some problems, it also opens the door for larger ones. The potential for misuse, even accidental, can affect every aspect of a person's life, unlike issues with physical documents or decentralised systems. Entrusting identity holistically to any system, even for protection, means relinquishing control.

My *Oma* died *with* her identity, not *because* of it. She was more than an elderly lady defined by her birthdate and address; these were mere details that travelled with her on the arc of her life, not the sum of her being. Yet these details profoundly influence our future, as our personal traits often dictate our destinies.

Identifying Human Life: A Practical Discussion

Despite this deeply philosophical discussion, the easiest way to refer to each other in daily life is through our personal characteristics. My neighbour calls me Paul, because that's how I've voluntarily identified myself to him, and it suits the purpose of our mutual exchange. He is satisfied it's enough to write on my birthday cards and get my attention over the fence. I am satisfied it's enough for him to know where to place my Amazon packages when they're mistakenly delivered to him.

Neighbours Over the Fence

Our neighbourly agreement inherently implies that the identity characteristics we share with each other are tightly scoped. That means we are to use them for limited purposes, trusting each other to respect these boundaries. We have built this trust over time, but certain situations may require us to revisit and renegotiate these terms.

He may know many people named Paul and might request my surname to distinguish me in his contacts list. Should I provide it, I'm aware that it alters our relationship; I will have shared my full legal name with someone who knows where I live. This adds a layer of risk, but also deepens our trust, as it shows my willingness to voluntarily share more of my valuable identity.

Out and About

In the physical world, we commonly identify each other by using context-specific characteristics, rather than just names. For example, my local barista knows me, not by my name, but as the early riser who orders double-shot long blacks and pays with cash. Our interaction is based on a socially assumed protocol: a brief exchange while he prepares my coffee, a chat about the weather, payment, and then we part ways.

In his book, *Learning Digital Identity*, Phillip Windley describes this phenomenon as tacit knowledge. He highlights our innate ability to recognise and interact with others based on limited information and implied social boundaries. (21 p. 19) This skill, like a sixth sense, underscores the inadequacy of centralised government identity management, particularly in the digital realm where precise modelling is required.

Aggregated Characteristics

Claiming an identity today goes beyond stating a name or coffee preference; it's about substituting the statement, "I am this person" with "I have these characteristics." Consider someone claiming to be King Arthur. If King Arthur is widely known for his ability to extract a broadsword from stone, then successfully doing so validates that someone is who they claimed to be. The more unique the identifying trait, the more confidence we have in the claimant's identity when demonstrated.

Types of Identity Characteristics

A human identity consists of many distinguishing characteristics, each with different levels of universality, stickiness, and uniqueness. For instance, a birthdate is universal and sticky; everyone has one, and it remains constant throughout one's life. However, given the limited number of days in a year, a birthdate is not a unique identifier, as it's shared with around 0.3 percent of the population.

A residential address is another identity characteristic; it is not universal as it doesn't apply to the homeless, and less sticky than birthdays, since it can change throughout one's life. However, addresses are more unique than birthdays, as it's rare for 0.3 percent of a population to share the same living space. Similar assessments of universality, stickiness, and uniqueness apply to other identity traits like sex, phone numbers, and club memberships.

Problems with Aggregating Identity Characteristics

Traditional centralised identity verification methods face challenges, due to the varying number and types of these characteristics.

Any conclusions drawn from them are essentially just estimations. Additionally, many of these characteristics, like phone numbers, are re-purposed; primarily intended for other uses and not ideal for identity verification. For example, I share my phone number with my neighbour to facilitate communication, not so he can identify me.

Human identity's complexity cannot be fully captured by systems that reduce it to a single identifier, like a serial number for equipment. Nor can it be represented by re-purposed characteristics, each one individually lacking the properties for unique identification. Yet this is exactly what governments attempt to do with digital ID.

Digital ID systems require the aggregation of countless personal traits to sufficiently estimate an identity.[29] In recognising the complexity of defining 'identity', we must be wary that such systems could pigeonhole our multifaceted selves, potentially dictating our access to services, and rights, based on a narrow subset of digital attributes. Moreover, there is no rigidly-defined rule for the exact amount of aggregation needed to identify someone. The criteria vary, as different interactions demand different levels of confidence.

The Runaway Demands of a Broken System

Society's current identification model has clearly reached its limits. It relies on a patchwork of personal characteristics, which has led to an over-complicated and ineffective system. A misguided solution to these problems is to do more of exactly what caused them in the first place. This is seen in the escalating requirements of Know Your Customer (KYC) and Anti Money Laundering (AML) regulations in many countries.

This approach demands that individuals disclose increasingly more personal details under the pretence of enhancing verification

---

[29]   We see this in the need to present multiple forms of identification in some instances, such as a driver's licence as well as utility bills, passports, or welfare cards.

confidence. Yet, each incremental scope increase only temporarily satisfies the system's needs. Like a drug addict seeking their next hit, inevitably further demands for identifying data will follow.

Persisting with ineffective methods is futile. For instance, the problems caused by fuelling a car with vinegar won't be resolved by adding more vinegar. Genuine solutions are fit for purpose, emphasising quality over quantity.

Individuals, as owners of their identity data, are uniquely qualified to decide what personal characteristics to share, and with whom. Rather than centralising all information with a government for potential future use, people should evaluate the necessity of each identity verification request: is it a genuine need, or a convenient want?

This personalised approach to identity verification, involving negotiation between the requester and provider, helps ensure that individuals can personally manage responsibility for their risk profile and decide when and how to disclose their identity information. We'll return to the topic of centralised decision-making in the next chapter on trust.

Verification Scope

What are the implications of needing to share different characteristics for each identity claim? The subjective nature of distinguishing the scenarios demanding robust evidence, and those that may be verified casually, risks oversharing personal information. This exposes people to undue dangers, especially under involuntary conditions that lack negotiation.

Normal Circumstances

Take a trivial example. Collecting a handwritten name tag at a child's birthday party is a simple act that doesn't require extensive documentation, like bank statements or utility bills. The party host, often a family member, is likely to trust attendees at their word without formal verification, given the existing relationship with the child. In this low-stakes context, the worst outcome of a verification error is a minor embarrassment, like a name mix-up during party games.

Contrastingly, using a handwritten name tag for customs clearance at an airport would likely lead to a back-room conversation. This situation demands a higher level of proof due to the serious implications of incorrectly identifying someone at immigration. Typically, multiple identity characteristics are required, with the accepted format limited to government-issued passports.

In the first scenario, party attendees and the host voluntarily negotiate name tag issuance. The second scenario mandates specific identity disclosure, potentially barring entry into a country for those uncomfortable with the process. The burden of involuntary identity verification is becoming more onerous, with the growing expectation for additional forms of identification suggesting that government-issued IDs are insufficient. In a twist that would have once been considered satirical, signing up for some financial accounts today requires a selfie with a card showing a unique code and date.[30]

The way we are asked to identify ourselves can even differ for identical purposes, as seen during the COVID era. For example, some public libraries only needed a simple hand-written sign-in for identity proof, akin to the birthday nametag scenario. In contrast, others required a driver's license, more like the customs scenario. Even within the same library, the verification scope could change based on the staff member present, leading to inconsistent practices, and highlighting the problems with using aggregated identity characteristics for verification, even before considering malicious intent.

---

[30]   I once had such an application rejected because my elbow was out of frame and the image hence did not clearly show that the hand holding the card was attached to my body. Can we agree that modern identity management is broken?

Abnormal Circumstances

These examples highlight how we're often expected to reveal aspects of our identity to unknown entities, who reciprocate at best with generic, often meaningless privacy policies. In calm periods, the risks of data sharing are not obvious, leading us to favour convenience over anonymity, such as with tap-and-go payments over cash. While the identity details exposed in these routine activities often exceed what's reasonable, many continue to accept this trade-off.

During crises, however, verification scope can quickly change, shortening the time between identity exposure and its impact on daily life. The COVID era showed people's willingness to surrender identity due to fear or perceived altruism, an almost overnight shift. This change in scope, requiring digital identification for previously unrestricted activities, down to entering buses and buildings, shows how identity can control access to resources in response to rapid policy changes, even those we may disagree with.

The adoption of QR codes and government tracking apps during this era, led to the extensive collection of detailed personal data. Post-crisis, this information remains stored and has already been inappropriately re-purposed by governments.[31] The ease with which a central authority can reuse digital identity data is a critical issue we will return to when discussing technology in Chapter Four.

Behavioural Conditioning

More disturbingly, the COVID era significantly altered our communal approach to identity, embedding a culture of mutual surveillance and incremental expansion of identity checks. Phrases

---

[31] In 2021, then Western Australian Attorney-General John Quigley revealed that police had issued multiple notices to the state health department "requiring them to hand over data from the SafeWA app since mandatory registration at venues was first enforced." This development was described at the time by the then opposition leader as a "massive breach of public trust" after previous government assurances that public COVID tracking data was solely for contact tracing purposes. (95)

like "Have you checked-in?" became commonplace, normalising the proliferation of identity verification points, where none existed before. This shift also led to a mistaken association of 'digital' with cleanliness and correctness, resulting in a preference for traceable digital transactions over more anonymous alternatives like cash payments and paper sign-ins. The New South Wales Council for Civil Liberties highlighted this in the 2024 Senate Economics Legislation Committee inquiry on the Australian Digital ID Bill 2024:

The COVID pandemic normalised the mandatory mass collection, use and storage of personal data using contact-tracing apps and as a result NSWCCL has increased concerns generally with ongoing data centralisation and datafication of Australians. (1 p. 83)

Identity Evolution and 'Scope Creep'

Even with a perfect identity management solution, challenges persist due to the constant evolution of our bodies and experiences over time. A 2022 report entitled *State of Digital Identity* recognises that "identities are constantly evolving." (22 p. 3). This evolution requires any identity registry to adapt, meaning the constant collection of more digital and behavioural data than initially anticipated.[32]

As our lives and identities continuously evolve, so too must the government systems that represent us. Left unchecked, this expansion, known as 'scope creep' inevitably leads to the complete exposure of all personal details globally, as there is no current mechanism to limit these escalating demands. Technology's rapid advancement has outstripped our capacity to manage how we interact with it, leaving behind an unwieldy trail of sensitive data that's difficult to audit,

---

[32] We will learn in Chapter Four that the Nazis withheld food rations as a means of ensuring continually updated identity records.

revoke, or secure. This disconnect reveals current thinking on digital ID as a square peg in a world of round holes.[33]

Unlocking Human Resources

Regardless of how we arrived at this situation, we must navigate the fact that identifying ourselves has become an integral part of modern life. Identity is deeply entrenched in the systems we rely on, leading many to question the importance of protecting our personal data at all. Indeed, if identification was merely a means of addressing our neighbours, concerns would be minimal. Beyond that, however, understanding the need to be cautious about who knows our identity requires recognising the human resources it unlocks.

Identity is not merely a means of referring to someone, it is a gateway to their assets and potential. As the drivers of technology and civilisation, humans are the ultimate resource. Knowledge of identity provides access to this invaluable productivity.

In *functional* societies, judiciously sharing identity is hence crucial for economic and social progress. For instance, disclosing educational qualifications and work history to potential employers can prove suitability for paid work. Likewise, sharing financial details with banks helps determine business loan eligibility. When *correctly* scoped and *managed*, identity lubricates the machinery of progress.

Conversely, in *dysfunctional* societies, the forced collection of identity becomes a tool for control. Knowledge of skills and behaviours can lead to exploitation through forced labour or blackmail. Likewise, the misuse of identifying characteristics used to verify property ownership can simplify and even encourage theft. When *incorrectly* scoped and *mismanaged*, identity incentivise the destruction of civilisation.

---

[33] For example, PayPal recently updated their terms of use resulting in a termination of service for those users unable or unwilling to provide additional identity verifying information such as a residential address.

The Local Context

While the connection between identity and resource access may seem abstract, it has profound implications at the local level. Identity's management—or mismanagement—can have immediate, tangible effects. During a recent bank visit, I overheard a teller asking an elderly woman for her name, phone number, and other personal details to verify her account. Like many daily interactions, this conversation was audible to everyone waiting in line.

Once authenticated, the clearly confused customer conversationally unveiled a cascade of additional personal information, including her son's name, job, employer, and work location. This moment of candour could have easily led to several issues: financial or identity theft, employment fraud, phishing, and social engineering attacks. [34, 35, 36, 37, 38]

The woman's account was vulnerable to those who overheard the identifying details used to unlock it, showcasing the power of identity in resource access. Such attacks are not uncommon, and can come from everyday opportunists, especially in times of high living costs when desperation is high. While the potential harm to an old lady at a bank is significant, national-scale risks dwarf it. A government collecting vast amounts of identifying data transforms into a silent observer, like a fly on the wall, privy to each and every transaction.

---

[34] If the hacker manages to gain access to the woman's online banking details, they could directly steal money from her account.

[35] The hacker could use the personal details of the elderly woman and her son to impersonate them. This could involve opening new bank accounts, applying for credit cards, or even taking out loans in their names.

[36] With the son's employment details, the hacker could impersonate him at his workplace or associated businesses, potentially gaining access to sensitive company information or resources.

[37] The hacker could use the information to craft personalised phishing emails or messages, tricking the woman or her son into revealing more sensitive information like passwords or credit card numbers.

[38] The hacker could use the information to manipulate customer service representatives at the bank or other institutions, convincing them to provide access to accounts or to change account details.

The National Context

The effects of identity mismanagement at a national level may not be as immediately apparent, making the link between identity and resource access less obvious. A poignant historical example is the Nazi occupation of the Netherlands, where the well-maintained Dutch identity registry, established for benign purposes, became a tool for oppression.[39] (4 p. 66) This prized spoil, containing a honeypot of sensitive data on religion, occupation, disabilities and more, allowed German forces to categorise and control the population as territories were annexed.

Misappropriation of assets by governments has not always constituted illegality. This is another key difference between identity mismanagement at the individual and national scale. A notable instance is the *1938 Ordinance on the Use of Jewish Assets* in Germany, which selectively curtailed property rights. Protecting property ownership meant withholding religious affiliation from government records.

Shortly after this law was introduced, those identified as Jews in the population registry were forced to "deposit all their stocks, shares, fixed-income securities and similar securities in a deposit at a foreign exchange bank." [40] The government was even allowed to sell Jewish businesses. (23 p. 71) Access to these resources required no less than approval by the Reich Minister for Economic Affairs. This effectively legalised the confiscation of private property from people with certain characteristics, identified by the population registry, and enforced by the police state. Such history underscores the

---

[39] This source of data is every bit a weakness of modern identity registries too. Identity databases are vulnerable to the will of conquering forces.

[40] Importantly, labels such as 'Jew' did not fall under well understood definitions. At the time, a Jew was considered to be someone with a genetic characteristic, not a practicing faith. Many people fell prey to discriminative laws having never practiced Judaism. In some cases, this was the result of the government having found evidence that a grandparent had been in some way associated with the label. Truly anyone could potentially be ensnared by such subjectively defined legislation, in some instances requiring only suspicion, making it incredibly dangerous and largely unavoidable.

dangerous relationship between identity management and asset control, at a national level.

The Competitive Context

Beyond malice, the use of identity to gain resources extends into competitive fields like chess, football, poker, and martial arts. Here, players meticulously analyse videos of their opponents' past performances to identify exploitable characteristics as weaknesses. Facing an unknown opposition levels the playing field, but having insights, such as a poker player's 'tell' or a fighter's 'favoured move', provides a significant edge. This leads to rewards like improved reputation and prize money.

Our identifying data is not trivial; it's a valuable asset. In the wrong hands, it can be nefariously traded among those who deal not in tangible goods, but in favours. Protecting our identity is crucial not just for safeguarding property, but for upholding human dignity. Centralised identity systems often fail to meet the standards required to preserve our humanity, and societies go awry quickly when that humanity becomes undervalued.

De-humanisation

While our competitive nature highlights the strategic use of identity to gain advantage, it also sets the stage for a far darker application: de-humanisation. In this chapter's final exploration, we unveil how the quest for advantage can devolve into a disregard for the person behind the identity. This phenomenon, like the trunk of a tree, grows steadily and is nurtured by its environment. Lesley Blume, the author of *Fallout*, an exposé of horrific government deception, articulates this chilling truth:

> *Once humans stop seeing other humans as human,*
> *the most horrific acts are possible. (24)*

De-humanisation is a psychological mechanism for viewing others as less than human. It facilitates atrocities by reducing empathy and moral restraint, making violence against perceived out-groups more justifiable. The process finds favour with those who see people as mere data points, prioritising utility over humanity. Once de-humanised, the downward-spiralling path is opened for governments to treat people in the same way that arborists treat trees.

De-humanisation's insidious nature subtly permeates our lives, evident in video games where players assault virtual characters without real-world consequences. Reflecting on my gaming days, I recall how easily pedestrians in *Grand Theft Auto* could be killed. Often, I would laugh at the absurdity of it. Gamers cheered as garbage trucks sent grandmothers flying. The more inventive the violence, the greater the applause.

These games not only simulate realistic human-on-human violence but also incentivise it, constantly rewarding players with dopamine hits for successful kills. The digital victims' facelessness serves to disconnect aggressors from the reality of their actions, making it easier to enjoy causing harm. This detachment, especially from the eyes and face, is a well-documented psychological adaptation, seen in the interrogation of hooded prisoners.

But what happens when this conditioning is paired with a politically charged, real-world environment of propaganda and fear? The Nazis found that de-humanising their victims made elimination efforts more efficient, with identity cards playing a crucial role. This deliberate process made mistreatment of people not only easy, but highly desirable.

Disposing of the Useless

Remaining vigilant for the early signs of de-humanisation in society is crucial. The process correlates with increased aggression and approval of discriminatory policies, as seen in contemporary examples of heightened support for bans against marginalised groups. The Nazis employed subtle tactics to initiate the de-humanisation process, beginning with the labelling and categorisation of people based on their identity characteristics, and the use of de-humanising language against those groups.

Propaganda was employed to mobilise the public against politically unfavourable groups. Rather than targeting individuals, they attacked concepts, normalising criticism of labels rather than the people behind them. These labels had slowly but steadily evolved over time, beginning with 'mentally unwell' or 'anti-social', and gradually stigmatising them as 'unproductive', 'needy', and 'useless'.

De-humanisation transcended social realms, enabling institutionalised categorisation. A German mathematics textbook exemplified this four years before the war. It depicted the societal cost of different groups of people using a curved graph, reducing human value to a mere mathematical equation. Embedding this thinking into the education system entrenched it at a societal level. (4 p. 95)

Far from just theory, the Reich Interior Ministry actually performed human value calculations. These quantified human worth through questionnaires, especially in nursing homes, comparing individuals' economic output to that of a 'healthy person', like valuing used cars against new ones. This process commodified human life, unfairly equating it to economic productivity.

Those who deviated too far from the benchmark, due to illness, disability, ethnicity, or misfortune, were categorised and treated accordingly. Some were neglected or denied medical care, while others were outright killed. In the three months from April to June 1941, the government murdered 29,200 people deemed economically unproductive, referring to these killings as 'disinfections'. (4 p. 97) These heartless acts were captured by correspondence from the Racial Political Office at the time:

Medical care must only be limited to preventing
the transmission of epidemics into the Reich territory.

We are not interested in the extent to which the population is pro-
vided with medical care by Polish or by the numerous
Jewish doctors …

… All measures that serve to restrict births must be tolerated or
encouraged. Abortion must be unpunished in the remaining area.
(20 p. 595)

Yet the regime sought influence on a grander scale. To that end, it implemented an innovated centralised identity registry and issued identity cards that we will explore in Chapter Four. People tend to trivialise cards, which can be lost, traded, or burned without loss of sleep. Nazi strategy directly linked the fates of these cards to the fates of the individuals they represented. By psychologically replacing humans with their ID cards, the task of eliminating people, like little Nadine, was made much easier.

The Nazis further de-humanised victims with the use of gas chambers. This method minimised psychological trauma for soldiers, who previously killed with ropes or bullets, exposing them to engorged eyes and bleeding brains, respectively. Sealed chambers distanced the perpetrators from these experiences and enhanced their efficiency to enable a predictable and measurable rate of execution. Gas chambers were the result of earlier pioneering attempts to distance the death merchants from the humanity of their victims. Such innovations included special vans that "pumped carbon monoxide into sealed passenger compartments." (25) This process efficiently asphyxiated people while in transit to mass graves.

Exploiting the Useful

Assigning the Blame

Economically productive people fared little better. Nazi dehumanisation had enabled the exploitation of forced labour in factories and camps. There, the regime's centralised knowledge of individual capabilities was traded with private industries willing to extract that labour to produce profitable goods for the war effort. This produced a symbiotic environment, described by Annie Jackson as one where:

> *Humans and machine parts went into the tunnels.*
> *Rockets and corpses came out. (5 p. 14)*

Beyond *IG Farben* (Nadine's executioner), various companies capitalised on this forced labour pool identified by the regime. For example, *Heinkel* used slave labour, specifically chosen from government records, for manufacturing military aircraft and advanced propulsion systems.[41] Likewise, *Steyr-Daimler-Puch* significantly contributed to the war effort by producing weapons, vehicles, and engines, almost completely reliant on coerced labour.

Steyr-Daimler-Puch holds the dubious honour of being the first arms corporation to employ Nazi slave labour, a decision surprisingly driven not by the Nazis, but rather by the company's director, Georg Meindl. Meindl actively lobbied senior officials for "early and preferential access to labour from concentration camps." (26). This enabled him to select workers with specific skills from the German population registry. Meindl's actions underscore the extent to which society was willing to commoditise its population and exploit

---

[41] *Heinkel* even produced jets and liquid-fuelled rockets.

human resources, unlocked by centralised knowledge of their characteristics.

Understanding the Reality

The haunting scenes that unfolded within the corporate production lines of that era mirrored those from ancient Egypt, with workers literally subjected to the crack of whips. Numbered people were at the mercy of their masters' ever-shifting whims. A simple misplaced gaze, deemed as insubordination, could swiftly lead to being "garrotted or hanged." (5 p. 14) This even took place on the factory floor, much like one might destroy faulty machinery in frustration.

Reflecting on Nazi Germany's identity management serves as a stark warning, illuminating the dangers of modern equivalents. A society that inventories its people risks de humanisation. In contrast, a society that resists such attempts acknowledges the unique human characteristics that exempt us from this treatment.

History's sobering lessons underscore the need to protect our humanity against the perceived utility of systems that reduce people to searchable, cross-referenced numbers and categories. Maintaining a clear distinction between object categorisation and human sanctity is crucial for any society aiming to transcend nature's raw predation. Given our historical struggle to uphold this distinction, however, we must prevent systems that enable de-humanisation from taking root.

Chapter Takeaways

Today, identity is almost as much a part of our online presence as it is our physical existence. It is hence crucial to be vigilant about protecting personal information both online and offline. Drawing from the lessons of history, we can take practical steps to help ensure our identities remain secure and reflective of our true selves.

In Private

Strategic Privacy: Simple Tactics for Protecting Your Identity

Be mindful of the identifying details you share, especially regarding your religion and profession. These aspects can reveal more about you than you may intend and could be used against you in unexpected ways. Treat your personal information like strategic assets in a game where you wouldn't want to reveal your vulnerabilities unnecessarily.

Becoming aware of what identifies you, like your voice, is a valuable personal exercise, allowing you to be conscious of how it is exposed. During voice-automated calls, consider altering your tone or using a disguised voice, which can prevent voice recognition from tying the call to your personal identity without hindering the system's functionality.

For everyday interactions like filling out unofficial online forms or placing restaurant orders, using a nickname can be a simple yet effective privacy measure. This isn't about deception. Using middle names or terms of endearment is common and offers a harmless way to protect your legal name until it's absolutely necessary to disclose, all while adding a touch of creativity to your interactions.

Remember, safeguarding your identity doesn't require drastic measures. The key is being conscious of the information and mannerisms you share, using discretion in everyday situations, and understanding the power of the details that make up who you are.

Guarding Our Children

To safeguard our children, we must be proactive about their digital presence. In an era where artificial intelligence can create digital twins using just photos and voices, protecting young digital identities is paramount. The story of poor Nadine, burdened by a pre-established presence, highlights the need for children to begin adulthood with a clean digital slate, free to shape their own identities and destinies without the constraints of a past they didn't choose.

The image of a child behind detention centre fencing in Sri Lanka, as seen by my in-laws, highlights the global vulnerability of children when identities are compromised. We must be mindful of the information we share about children, even seemingly innocuous details about their lives. These are not ours to share. As parents, we are simply guardians of them … until they are adults.

Discussing digital privacy early can empower children to manage their digital presence wisely. As parents, we have a responsibility to adopt this role for our children rather than assume that their teachers or peers will. Practical steps like setting up social media privacy controls and teaching the importance of consent before sharing personal information can protect them. These measures help children shape their identities on their own terms, free from digital shadows.

The Devil's Bargain: Navigating Promises of Future Benefits

The idea that sacrificing our freedom today will ensure greater security tomorrow is flawed and contradicts our cultural narratives. For example, the biblical story of the Israelites, led by Moses out of centuries of Egyptian bondage, illustrates that true safety and security are achieved through the unwavering affirmation of freedom, not its surrender.

Freedom and security do not come at the cost of exposing our identity. To wisely navigate such propositions, a healthy dose of scepticism is always warranted. Remember, preservation of freedom is the cornerstone of security and prosperity. By asserting our liberties today, we lay the groundwork for a future that is both secure and free.

Handling Biometric Data with Care

The convenience and efficiency of biometric technologies like facial recognition and fingerprint scanning are compelling. However, the significant risks they pose become apparent upon closer examination. Biometric identifiers are immutable—they can never be changed. Unlike passwords, if exposed, your biometric data could irreversibly haunt you throughout your life, acting not as a helpful tool but as a persistent ghost. Use strong, unique passwords managed by a reliable password manager to avoid the irreversible risks of biometric data.

Limiting Data Sharing

Exposing personal information today is sadly commonplace. Exercising caution when sharing email addresses, phone numbers, or birth dates is increasingly important. Pause to assess the need and trustworthiness of anyone requesting these, as each detail divulged, can increase the risk of identity theft and fraud.

Adopting a guarded mindset helps you identify and secure potential vulnerabilities where personal information could be exposed. For example, avoid leaving documents with your name and address visible in your car to deter identity thieves. Likewise, be cautious with unsolicited emails or messages that request personal information. Using a Postal Office (PO) Box for your mail delivery can also reduce the exposure of your identifying details to prying eyes. These preventive measures are not only legal but commonly recommended.

We can engage in thoughtful negotiations when asked for personal details. For example, when my insurance company called about a claim, I declined to provide information since I had not initiated the call. Instead, I proposed calling back using the official number from their website to confirm the caller's legitimacy before discussing details of my claim. This method is generally well-received and protects against impostors, who often pressure for immediate information.

By fostering a culture of vigilance and informed caution, we can collectively enhance our defence against the ever-evolving landscape of digital threats, especially those originating from government.

In Public

Control Your Personal Information

Beyond individual actions, we must collectively advocate for privacy rights and ethical identity management. The scenario of differing COVID sign-in processes at libraries highlighted how personal information requests can be unnecessary or arbitrary. Simply questioning the need for such requests— "Why is that necessary?"—can often lead to their withdrawal.

Consider a personal experience at a juice store where I was asked to fill out a digital form on a tablet just to buy a drink. It needed my address, name, and phone number, all unjustifiable for a simple purchase. By opting to leave rather than comply, customers can signal to businesses that such intrusive practices are unacceptable, encouraging them to reevaluate their data collection methods.

Being comfortable with questioning identification requests is important. Remember, justifications and intent can change. The repurposing of COVID check-in app data demonstrates that entities, especially governments, may not always adhere to their initial promises regarding the use of your data. Be mindful that once shared, your information could be used in ways beyond your initial consent. Before divulging details, think about whether you would be at ease with that information becoming public.

When the time comes to share your identity, speak quietly, unlike the elderly lady I overheard with the bank teller; or write it down for them to copy. Doing so emphasises the information's sensitivity and encourages careful handling by the listener.

Educate Yourself and Others

Educate yourself and others about the risks of digital ID in government hands and the critical importance of privacy. Leveraging your unique strengths is crucial in advocating for information privacy. Whether you're a public speaker, teacher, or parent, utilise your skills to promote the significance of protecting personal data. Not everyone needs to fight the same battles, but everyone can make a meaningful contribution in their own way. By implementing these takeaways, we not only protect ourselves but also actively contribute to a future that preserves the essence of our humanity.

Redefining Identity: A Positive Message

As we close this chapter, we can now see identity not as a document, but as the core of our shared humanity. Protecting it is essential to maintaining dignity. The issues dissected in this chapter teach us that unchecked identity control can unravel the very fabric of society. Digital ID is hence not a new concern, but rather a continuation of this long-standing narrative.

Outdated thinking and the unreliable judgments of those in power plague modern identification systems, which often reduce people to mere entries in a government ledger. Why do we persist with such methods? Have we learned from historical identity management mistakes? Are current efforts enough to prevent reoccurrences?

Our unique and unpredictable qualities distinguish us from mere data points. The broad application of government digital ID, a system defined by data points, risks creating more problems than it purports to solve. We must transcend bureaucratic centralisation, which has time and again shown its vulnerability to abuse.

There is hope, though, for the future remains unwritten. Our resilience and creativity have consistently surmounted past challenges, and they can do so again. A future beckons, where technology allows people to share identity at their sole discretion. In the next two chapters, we will explore viable solutions that transform privacy and consent from ideals into accessible realities.

Our goal is to foster a culture that values these qualities—not because we have something to hide, but rather something to protect. In the end, the measure of our progress is not the sophistication of our technologies, but how much they reflect the best of what it means to be human. As we turn the page, let's commit to a better future by considering the question that government digital ID claims to answer: how do we trust each other?

CHAPTER 3:
Trust

It is clear from the many hundreds of individual submissions that have expressed serious concerns about the privacy implications of Digital ID, that the Government has not yet earned this trust.

—Australian Senator David Shoebridge, Senate Economics Legislation Committee Dissenting Report, commenting on the Australian Digital ID Bill 2024. (1 p. 79)

At the moment our society, by and large, operates on the basis of trust. Now and then people are asked to identify themselves. It is the Commission's expectation that once a universal identity card system is established, that order will be reversed.

—His Grace, the Right Reverend Michael Challen, Bishop of the Anglican Diocese of Perth, Australia, and Chairman of its Social Responsibilities Commission, contributing to the Joint Select Committee inquiry on the Australia Card. (12 p. 178).

So much depends on reputation – guard it with your life.

—Robert Greene, *The 48 Laws of Power*. (27 p. 37)

One can argue that the feeling that one can develop with the card is that everybody is guilty and the only way you prove your innocence is by the production of cards when you are doing your dealings.

—Australian Labor Party MP, Mr John Saunderson, contributing to the Joint Select Committee inquiry on the Australia Card. (12 p. 179)

Strangers at a Military Ball

After a war that scarred landscapes and souls alike, my grandparents embarked on a remarkable journey. This chapter explores the foundation of their quest, peeling back the layers of identity to reveal the driving force behind it—trust.

The Holocaust had exposed the devastating effects of extreme identity commodification. Catastrophic government actions had burned civilisation to ashes, yet from these embers arose new opportunities to re-discover, and appreciate, the elegance of direct relationships.

My grandparents were at the front line of this re-discovery. From their inaugural dance at the *Vienenburg* ball, they cultivated an organic bond. As strangers, they embraced vulnerability, shared their languages and cultures, and allowed their love to heal the hatred that had coursed through Europe's veins for decades. Their investment sheds light on the workings of trust, from personal to institutional levels, offering lessons that echo into the digital age.

Unique Birthplace: A Symbol of Union

My grandfather, Bill, remained stationed in Germany after the war, his duty shifting from combat to care. During this time, he married my *Oma*, a union providing solace, amidst the grief of her younger brother's untimely death. Their shared optimism culminated in the birth of my father in 1950, at a British Army on The Rhine (BAOR) hospital, named *BMH Rinteln* (Figure 8 and Figure 9).

BMH Rinteln, one of several BAOR hospitals in Germany at the time, was regarded as British territory, much like an embassy. This unique status bestowed a poetic nuance upon my father's birth: he took his first breath as a German-born, British citizen, in a sanctuary of healing and hope.[42] His birth symbolised the

---

[42] The humanity displayed at *BHM Rinteln* at the time was evident in the details. One British major for example would bring smiles to children's faces by waking them at five o'clock to feed deer through ward windows. (94)

harmonious blend of his parents' backgrounds, and the enduring legacy of their love.

Figure 8. BMH Rinteln—the German hospital where my father was born. Picture scanned from slides and shared graciously by Philip Basford.

Figure 9. The beautifully manicured grounds of the BMH Rinteln Other Ranks (OR) quarters on Waldkaterallee (Forest tomcat avenue) circa late 1950s. Courtesy of Philip Basford.

Born into the 'boomer' generation, my father entered a world that was free from the harsh scrutiny of centralised power. Through the blood spilled by others, he inherited a time when humanity's uniqueness, like spring flowers, blossomed anew. His birth marked a shift where people were starting to learn to trust each other again, based on the merits of character, over the markings on a card.

The Paradox of Trust

In this crucible of German history, my *Oma* shaped her enduring view on trust. Her insights, a time capsule from that era, reflected hard-earned lessons. These emphasised the risks of large-scale governments, dictating the dynamics of small-scale interactions. Her social conduct, deeply influenced by her childhood, showcased the complex nature of how humans relate.

Paradoxically, my *Oma* was at the same time one of the most, and also one of the least trusting individuals I knew. She was highly trustful of her close-knit circle, family, and friends, with whom she had meticulously built social capital over time. Yet, her demeanour in situations demanding more intimacy than warranted was a study in contrasts. Something to behold, these encounters were often met with silence, and protective body language like an averted gaze.

During my dating years, I often disagreed with my *Oma's* stance on family inclusivity. While I saw the title of 'my girlfriend' as an automatic ticket to family events, she considered relationships a gradual journey, where the privileges of increasing social familiarity needed to be earned over time.

Young and ignorant, I was offended by her refusal to respect a label I assigned and revoked at will, a label that dictated who could enter her home—and life—on a day-to-day basis. By the same token, *I* always had a seat at her table, despite my sometimes-undeserving conduct. Our differing views sometimes led to conflicts, but I can now appreciate her principled approach, as it offered me valuable insights into relationships.

Even in her later years, distanced from her past, my *Oma* maintained strict discipline and discernment in protecting the space around her identity. As her grandson, I struggled to comprehend this aspect of her character. One telling event occurred at a local fair we attended, when we encountered a stranger of her age, also from Germany. As the pleasant conversation turned more personal, my *Oma* ceased talking, without even acknowledging their common birthplace. Baffled, I courteously ended the conversation.

These juxtapositions invite reflection. How can people exhibit openness and trust in some situations, yet be guarded and distrustful in others? Moreover, how does one become close to anyone without being given an opportunity first? Trust is clearly a complicated and complex issue, one that requires careful consideration before being invoked as a rationale for identity registration.

In a society that relies on personal identification to establish trust, one must question whether there are better tools for the job. The paradox of trust challenges the *'one-size-fits-all'* approach favoured by centralised digital ID systems. Unpacking these issues calls for a deeper look at the nature of trust and questions the effectiveness of using identity in promoting true social cohesion.

Trust: A Nuanced Examination

Trust varies with time and across relationships. Each scenario—be it with partners, work colleagues, or governments—demands distinctly different acceptable behaviours. To understand the shortcomings of identity registries, we must examine how information sharing builds trust, and how this dynamic is influenced by coercion and centralisation.

The Physics of Trust

Trust is like gravity—it's an ever-present force, even if we don't see it. It holds our relationships and communities together, much like gravity keeps our feet on the ground. Just as scientists like Sir Isaac Newton explained how gravity works, helping us to send satellites into space, figuring out how trust works can give us valuable insights into our personal connections, and how society should function.

Think about how we understand the world around us. In physics, we have two main theories: one describes the large scale, like planets and stars (classical mechanics), and another explains the small scale, like atoms and particles (quantum mechanics). Even though these two views are very different, and even contradict each other, both are validated within their respective domains. For example, classical mechanics accurately predicts the arc of a thrown ball, while quantum mechanics explains the strange behaviour of particles that are too small to see.

This is similar to how trust works. The trust between two friends is different from the trust we place in institutions or whole countries. However, we often use the word 'trust' for all these different situations, which can be confusing. This is like trying to use the rules for throwing a ball, to understand how an atom works—it doesn't fit.

We need to be clearer about what we mean by trust in different situations. Just as physicists use specific terms for different scales, we should develop better ways to talk about trust. This would help us build stronger relationships and better societies, knowing exactly what kind of trust we are dealing with.

Relational Trust

Let's introduce the term 'relational trust', as an analogy to quantum mechanics in physics, to describe the nature of human connections at a small-group, or atomic level. This concept, inspired by scholars like Harvard's Robert D. Putnam, a distinguished political scientist known for his work on social capital and community dynamics, emphasises the importance of mutual investments and voluntary actions in building trust within personal relationships. Relational trust is the cornerstone of human connections and captures the essence of intimate associations, shedding light on the *warmer* side of my *Oma's* character.

Relational trust flourishes in environments characterised by shared values, time investment, and 'skin-in-the-game' experiences, operating independently of external legal systems. This human-centric framework, if influenced by external laws, would see its motivations and outcomes skewed by a desire to avoid external punishments, rendering it involuntary, and undermining its relational nature. To clarify, it is useful to briefly demonstrate how relational trust operates in real-world scenarios.

Consider my neighbour, introduced earlier. Our trust evolved over time through small acts of integrity. I lent him tools without requiring government ID or a contract, and he reciprocated by returning them promptly and in good condition, even voluntarily expressing appreciation with a bottle of wine. Through these exchanges, I become increasingly comfortable lending items of larger value.

Equally, I don't depend on government laws to safeguard my wallet from theft at a family member's home either. The relational trust I have built through positive interactions and mutual accountability lets me safely leave this valuable item on the kitchen bench. In such cases, I rely on relationships over regulations, and prevention over punishment. These examples show that relational trust is built on kindness and mutual respect between known parties.

Enter the Accountability System

Relational trust can further extend to the immediate community, where a common environment encourages good behaviour. For example, I abstain from car theft not only because I believe it to be wrong but also to maintain a safe neighbourhood, knowing that such behaviour invites reciprocity. Put simply, I don't want to live in a suburb where cars are stolen. My decision is rooted in both a universal moral stance against theft and a vested interest in my local community's well-being, making government laws against car theft somewhat superfluous in this case.[43]

Clearly, relational trust works when actions and consequences are closely aligned, as captured by the adage *'don't spoil your own back-yard'*. However, this principle becomes less applicable as we broaden our perspective, as evidenced by the need to keep wallets secured in public venues. If relational trust was universal, there would be no need for such security measures, and we could leave our wallets lying around without fear. Yet, the reality is that a world without measures to protect private property remains an unattainable ideal.

Institutional Trust

As our social circles expand beyond small communities, relational trust, like quantum physics, slowly starts to break down. At the level of corporations, governments, and nations, history has shown count-less times that individuals in power can act without facing personal repercussions. This disconnect was what allowed Georg Meindl to safely broker a slave labour deal between *Steyr-Daimler-Puch*, and the Nazi Government, clearly demonstrating how those removed from the consequences of their decisions can *'mess up other people's backyards'*. This necessitates additional mechanisms in systems designed to establish trust at scale, helping to ensure accountability and prevent misuse.

How do we expand relational trust? Let's introduce the term 'institutional trust', as an analogy to classical mechanics in phys-ics, describing the nature of human connections at large scale.

---

[43] I don't steal cars when I'm travelling either, for example, and still wouldn't if it was legal.

This concept, aligned with thinkers like Francis Fukuyama, a senior fellow at Stanford renowned for his work on political development and international politics, emphasises the importance of implementing protocols in software or contracts to establish and build trust between larger parties unknown or unrelated to each other. Institutional trust captures the inherent complexities when interactions exceed the scope of immediate, personal relationships, shedding light on the *colder* side of my *Oma's* character.

Trust and coercion are mutually exclusive. To effectively estimate trustworthiness on a large scale, systems must operate without the immediate feedback, local knowledge, and reputation that inform trust in smaller communities. However, protocols that rely on coercion or forced participation, like government mandated digital ID, distort the system's ability to reliably assess trust. Coercive systems undermine voluntary interaction and are prone to corruption, stifling competition, and choice.

Enter the Legal System

Legal systems across the globe are specifically designed to enforce centrally-determined standards. This largely serves to implement institutional trust but does so poorly. It is a reactive system, usually accessible only to those with financial means, which focuses on punitive cure rather than encouraging prevention. Moreover, laws which are the domain of governments, are limited in their ability to regulate government actions themselves, especially in the conflicting presence of political donations and kickbacks.

The legal system becomes a game where the rule-makers are also players, with the power to alter the rules if required to ensure victory. Moreover, laws impact only those who choose to follow them, making them imperfect safeguards. As a result, we will always need pockets for our wallets and identity management systems with similar security measures.

Although most people do not steal wallets, and most governments do not act like the Nazis, it only takes one to cause significant harm. Yet, unlike the relatively simple issue of replacing a stolen wallet, the consequences are far greater when we consider the realms of identity and governance.

When Relational and Institutional Trust Collide

Healthy societies understand the distinction between relational and institutional trust, whereas unhealthy societies often blur these lines. For example, relational trust can be undermined when cultural cohesion within communities diminishes, allowing institutional trust to fill the void and be imposed inappropriately. This can occur during periods of societal stress, such as during the COVID era, when families were encouraged by governments to distance themselves from their own members and even rescind Christmas invitations based on their medical status.

The danger intensifies when these different trust domains are deliberately conflated. For example, governments could potentially benefit by convincing citizens to view their institutional digital ID with the same level of personal engagement and trust that is typically reserved for relational interactions. This scenario is akin to treating unknown government officials as if they were close personal friends with a shared history.

Severe consequences can result when the lines between relational and institutional trust are blurred. History provides numerous cautionary tales of excessive punishments enforced by institutional systems, in the name of community welfare at the local level, such as in early Nazi Germany. Therefore, we must maintain the distinction between relational and institutional trust. An easy test in any scenario is to ask yourself: *Is this the government's business?*

Trust at Scale: Different Approaches to Institutional Trust

People are sceptical of strangers. The challenge of establishing institutional trust arises from this inherent scepticism towards unknown people, especially powerful groups of them in government or business. Therefore, we cannot directly depend on these powerful groups to effectively safeguard us from their own potential misconduct.

The modern digital landscape often requires individuals to interact with entities without pre-existing relationships or trusted referrals. These interactions typically involve granting access, verifying claims, or confirming ownership, with identity serving as an imperfect means to this end. The fundamental challenge lies in establishing trust between two parties unfamiliar with each other, sufficient for the exchange of goods, services, or information. This centuries-old dilemma has puzzled minds throughout history.

The Byzantine General's Problem

To illustrate the challenge, let's delve into the Byzantine General's Problem, a concept rooted in game theory and battle tactics dating back to the fifth century Byzantine Empire. This scenario underscores the importance of trusted communication among distant military units in winning battles, which mirrors the trust required in today's modern online digital interactions. How can a general in one battlefield trust that the messages he receives are legitimate instructions from a fellow general in another battlefield, rather than an enemy impersonation? Coordinating synchronised attacks demands authenticity in communications. Verifying the sender's validity is crucial to determine which messages require action.

The general solution to this problem involves devising methods for each party to independently verify the claims made by the others. Numerous possibilities have been proposed by mathematicians, each with its compromises. One approach involves sidestepping the

issue entirely by introducing an additional, implicitly trusted, centralised authority to relay communications; let's call it the Byzantine
Command Centre. In this scenario, Byzantine generals in the field
need not fret about the accuracy of their communications, if messages brokered by this god-like authority are assumed to be inherently true.

Yet, this does not truly solve the problem, it simply shifts the burden of trust from the individual generals to the command centre
itself. The Byzantine Command Centre is, after all, staffed by
groups of implicitly untrustworthy people that we purportedly need
the authority to vouch for. Its coveted position of being beyond
question instantly makes it extremely susceptible to infiltration,
fraud, and corruption. The command centre claims to be trustworthy, even when the parties relying on that trust cannot audit, or
otherwise verify, the ongoing justification for it. In this situation,
trust exists simply because it existed before. Is this a system of
trust, or of faith?

While the centralised authority approach to the Byzantine General's
Problem may be functional to some extent for selected activities, it
nonetheless suffers major limitations. Importantly, it replaces the
challenge of achieving consensus among multiple parties, with the
risk of a single point of failure. Therefore, the true solution lies not
in centralising decision-making, but in developing protocols that
allow consensus, despite the potential for dishonest or unreliable
participants, as these factors are present in any system.[44]

Centralised Institutional Trust

Theology: Centralised Authority as a Shortcut

Despite its flaws, delegating large-scale problems to central authorities is easy. This workaround is hence common in various aspects

---

[44] Engineers refer to systems that operate correctly despite the presence of bad actors as
'Byzantine fault tolerant'.

of life. Renowned scientist and atheist Richard Dawkins, in *The God Delusion*, criticises this dependence on a centralised authority in theology, coining it *'the God of the gaps'*. His thesis likens God to an implicitly-trusted central figure that theists use to fill gaps in their understanding of the universe, similar to how a communication broker offers centralised clarity about the sender's identity in military communications.

Dawkins contends that using God as a stopgap for scientifically unexplained phenomena creates more issues than it solves. He criticises the God of the gaps approach for substituting genuine inquiry with a divine placeholder, arguing that unravelling human origins, despite its challenges, is simpler than attributing them to God, and then rationalising God's existence. (28) This mirrors communication issues, where directly verifying the authenticity of claims made by an individual is easier than justifying why authorities are trusted with doing so on their behalf.

Finance: Scope Creep on Display

The banking sector further exemplifies our deep-rooted reliance on centralised institutional trust. Banks mediate global transactions between strangers. These behemoths, heavily dependent on identity checks, must contend with all the issues of centralisation and scope creep already discussed. Yet originally, banking was far from complicated, emerging purely in response to a basic need.

In simpler times, a bank was a trusted tribe member, safeguarding a hunter's purse before venturing into the forest. This system relied on relational trust and clear consequences. However, as banking expanded worldwide, it maintained its trustworthy label, despite moving beyond the boundaries of personal relationships. The evolution to institutional banking did not fully account for this shift in scale, as we often build on what we know, without adapting to new contexts.

Lingering Issues with Centralised Institutional Trust

Being a trusted intermediary grants modern banks significant power, allowing them to freeze or close accounts at their discretion, leading to severe economic impacts for users.[45] This decision-making at scale, where people make choices for individuals they don't know, contrasts with the direct repercussions faced by a misbehaving hunters' bank.

Moreover, the widespread use of these juggernauts leads to extensive data collection and comprehensive identity profiling of customer transactions. This constitutes a covert breach of privacy, exposing users to market manipulation, collusion, fraud, and the fallout from potential bank failures.[46] Additionally, unauthorised access to this data, or 'hacking', often results in consequences more severe than those these systems aim to mitigate. These may include those invited by the talkative lady I overheard at the bank teller, as recounted in Chapter Two.

Beyond these risks, finance also highlights the tension of competing priorities. Fund transfers are fundamental to life, prompting consideration of whether urgent needs take precedence over privacy concerns. Does transferring funds to support a relative in need outweigh any potential data loss in doing so? Could there be methods to meet these critical global needs, without having to sacrifice anonymity by placing faith in large organisations?

Decentralised Institutional Trust

The Evolution from Central Authority to Trustless Systems

A 'trustless' system secures online transactions by having everyone adhere to rules established by a computer program, negating the need to depend on any individual or organisation to be honest or fair.

---

[45] The practice of 'de-banking' is occurring globally today.

[46] The collapse of Silvergate Bank, Silicone Valley Bank, Signature Bank and others, in March of 2023, highlights the very real consequences of instability in the space of centralised banking.

Trustless systems offer a real solution to the Byzantine General's Problem, ironically fostering more trust than traditionally trusted counterparts.

Trustless systems replace faith in centralised authority with decentralised verification and community consensus. These are driven by auditable algorithms, or decisions based on pre-defined and public criteria. Unlike legal frameworks, trustless systems ensure a game's rules are established before play begins. Adopters of this collective agreement method benefit from its universal verifiability and enforceability, allowing the public to independently confirm information and collectively exclude deceivers, all based on the unique, anonymous attributes of the participants.

Royalty

How can trustless systems translate to real-world scenarios? Let's return to our King Arthur story for a fitting analogy. The act of extracting the sword from the stone was the sole test to determine the true heir to the throne, not the individual's name or personal details. The phrase, "He who pulls the sword from stone, may sit on the throne," encapsulates this idea. It avoids the need for identity verification, which would complicate the process with an additional step, as in "He who pulls the sword from stone is called Arthur, and Arthur may sit on the throne."

King Arthur's legitimacy was not proven by a government identification document, but by demonstrating the publicly-recognised condition for kingship. Observers could independently confirm his qualification without relying on trust—they saw it with their own eyes. Similarly, decentralised consensus methods solve the issues of trust and identity by replacing blind institutional faith with verifiable, digitally distributed conclusions. Could it be worth adopting similar approaches in more realistic scenarios?

Finance and Beyond—Don't Trust, Verify

Decentralised financial security is not a futuristic concept, but a reality since 2008 with the birth of Bitcoin. Bitcoin is a trustless digital currency that operates independently of a central authority, using a computer algorithm, or predefined rules, to secure transactions and control the creation of new units. It allows users to reliably send or receive value over the internet directly to each other, quickly, cheaply, and reliably, without the need for intermediaries like banks. This development birthed the security mantra:

"Don't trust, verify."

Bitcoin uses a technology called 'block-chain' that implements the equivalent of observing a sword being pulled from stone. Its decentralised model bypasses the identification process, making the identity of account holders irrelevant by directly verifying account balances. This approach avoids the complications of identity verification, while providing a fully functional global money.

The potential societal benefits of this technology are profound. Imagine reliable, immediate, and auditable elections that don't compromise voter anonymity or rely on fallible electoral institutions. This could potentially lead to a shift from traditional general elections to micro-issue voting, where specific issues are open to real-time, secure public feedback. Using technology to arbitrate institutional trust both safeguards identity, and also challenges the necessity of traditional governance by enabling direct, distributed decision-making.

The Dual Nature of Building Trust

This discussion highlights the dual aspects of trust: asserting a claim, and verifying it. Whether it's my neighbour promising to return borrowed tools, generals coordinating an attack, or verifying financial balances, each claim has a specific goal: task completion, military success, or commercial transactions. Do we need identity proofs to validate these claims? Are there other, safer methods?

Anonymous verifications are effective for building both relational and institutional trust. Marriages, for instance, start with two strangers gradually learning about each other, and grow into one of the most trusted bonds humans can experience. First dates are an implicit claim that each party is who they say they are, and they are worth getting to know. Yet relationships don't start with the couple exchanging passports over the dinner table to verify this. Separating who we are from what identifies us should be our favoured way to interact.

Why are we so obsessed with identity? Why do governments continually tout their digital ID products as the golden standard for building trust? This gargantuan assertion unravels quickly upon scrutiny, as trust requires choice and cannot be forced. Trust must emerge from organic interactions, rather than from mandatory identity checks that reduce people to mere labels. Let's travel back to Germany to understand the repercussions of this reality.

Dissenters and Criminals: The Enigma of Labels

During one of my visits to Freiburg, Germany, I learned the phrase, *'Im Eimer'*. This translates to *in the bucket*. The phrase made sense when I learned that it means exhaustion, similar to the English expression for death—kicking the bucket. Clearly, language can be imprecise, and often deliberately ambiguous. Effective communication requires both good intentions and an understanding of social nuance beyond literal interpretations.

Discussing trust requires similar precision, particularly when used to justify government-imposed digital ID. Yet, we often encounter silence, censorship, or misdirection instead. As the saying goes, the devil is in the details—a lesson well-known to anyone familiar with product warranties. How can governments introduce such civilisation changing infrastructure on vague promises like "catching criminals?"

The allure of identifying threats is undeniable. Yet, it remains unclear how we can trust governments to do so, and who their potential targets are. We must pause to critically examine the subjective nature of terms like 'terrorist' and 'criminal', acknowledging the apparent dangers attributable to their over-simplification, not to mention the potential misuse of broad and subjective classifications in online identification. Are all Christians the same? Are all Caucasians the same?

Defining the Criminals

"One man's terrorist is another man's freedom fighter" according to Raoul, cigar factory manager and British sleeper agent in the James Bond film, *Die Another Day*. (29) While the distinction between heroes and villains in Bond films is quite clear, real life is filled with ambiguity. Take, for example, the controversy and longevity of the War on Terror, which stem, in part, from the seemingly arbitrary application of the term 'terrorist'.

The broad nature of terror allows for its potential and convenient re-purposing against any dissenting individual or group. Likewise, this ambiguity raises questions about the true targets of government digital ID initiatives aimed at catching so-called criminals. Who exactly are these criminals? And what crimes are they perpetuating?

Are we hunting musicians using marijuana for inspiration, or veterans using it for pain relief? Perhaps we should target peace activists protesting war, or patriotic supporters of troops? What about environmentalists fighting deforestation, or others supporting it, to build solar farms? The inherent risk is that terms used to justify punitive government systems can easily apply to anyone deviating from the mainstream narrative based on subjective judgments.

This risk was clearly demonstrated in Nazi Germany. Back then, the term 'anti-social' was initially used to stigmatise and shun political opponents. Yet those branded with this label were eventually forced to wear an identifying black triangle in *Auschwitz-Birkenau*. (30) This resulted from expanding the meaning of the word to encompass vagrancy, prostitution, and:

> *... a wide range of other deeds or behaviours,*
> *loosely and arbitrarily interpreted by the police. (30)*

Developing identification systems with punitive components risks ensnaring innocent people in a broad dragnet. The underlying common assumption is that such measures target 'others'—those with views divergent from ours. This prompts the crucial question: Who determines who the targets are?

Who Gets to Decide?

The Democratic Process

Who decides which groups digital ID will marginalise? The democratic process seeks to address this question, intended to give citizens a voice in shaping laws. Despite its imperfections, voting can help prevent a single viewpoint from dominating these definitions.

Free debate is vital for this process to function. With precise definitions crucial to how identity systems are used, we must regularly discuss who is considered a threat as societal norms evolve. Yet with new laws curtailing speech, is the power to define terms shifting from the community to those holding political office?[47]

---

[47] An example is *The Criminal Code (Serious Vilification and Hate Crimes) and Other Legislation Amendment Bill 2023*, Queensland, Australia.

Hate Speech

Nazi Germany again offers insights. The now common term 'hate speech' arguably even originated there. Similar labels were used then, just as they are now, as a controlling method to manipulate the definition of words, suppressing dissenting voices and differing opinions. One example comes from a 1938 report from the Reich Security Main Office which condemned churches for using language defined as harmful to the regime:

> *With whispered propaganda and inflammatory reports, church groups of the most diverse tendencies tried to bring insecurity and nervousness into the people's hearts and thus to weaken the foreign policy momentum of the Third Reich. (23 p. 93)*

Originally labelled as hate speech, is this now recognised as a brave stand against injustice? Today, few would oppose efforts that would have weakened the Third Reich's foreign policy momentum. Weaponising labels to quell dissent risks running afoul of the adage— never grant anyone political power you wouldn't trust in the hands of your worst enemy—for eventually, they will acquire it, and wield it against you.

Evolving Definitions: An Impossible Task

Defining 'misconduct' has become increasingly complex in the digital era. Automated systems designed to identify wrongdoers often fail to consider contextual nuances. Digital ID is particularly susceptible to these challenges, as it applies fixed conclusions to variable scenarios. As society and technology progress, our interpretations of right and wrong, and their enforcement, must evolve accordingly.

The Modern Context

Consider speeding. Speed limits were originally policed by humans targeting primitive vehicles on dirt roads shared with horses.

Although designed to mitigate risks from a bygone era, they continue to be enforced, now by unmanned digital cameras making binary decisions targeting modern, even self-driving cars. Consider also the allocation of blame, a once simple task. Should the responsibility for speeding today lie with the vehicle owner, or with the engineers who designed its autopilot software?

The Historical Context

In many historical contexts, the fluidity of definitions is starkly illustrated. For instance, during the Nazi regime, the classification of individuals based on racial purity exemplified the perilous nature of ambiguous labels. A notable example of this is a statement by the Reich Propaganda Ministry on March 10, 1943, where State Secretary Gutterer conveyed that Hitler had assessed Ukrainians as:

> *... in good order from a blood point of view and that there were therefore no concerns if parts of them were absorbed into our people. (31 p. 11)*

This perspective, however, was not consistently held. Hitler himself later expressed disagreement with this blanket acceptance, despite previously acknowledging the presence of German blood in Ukraine. This inconsistency highlights the inherent danger in attempting to rigidly define groups or individuals based on arbitrary identity criteria.

Historical precedents serve as cautionary tales for our current efforts to delineate wrongdoing, especially in the digital age. The attempt to categorise individuals or actions as definitively right or wrong, without due consideration of the complexity and variability of humanity, can obscure true intentions and hinder societal progress. The lesson here is clear: even with the most precise definitions, the act of excluding or penalising dissenters does not necessarily equate to an advancement in society.

Healthy Societies Need Dissenters

Why would we want to target people with opposing views? Dissenters play a crucial role in societal progress. Where would Western civilisation be without people like William Wilberforce, or George Washington? Individuals who thought differently and fought to sell their ideas to the mainstream. Digital ID systems could easily hinder such progress by unfairly targeting, and penalising innovators like these.

If Wilberforce had been subjected to a restrictive digital ID system during his life-long campaign against the British slave trade, would it have labelled him a criminal for his dissenting and economically destructive views? Would it have hindered his efforts by denying resources and liberty? This is food for thought and highlights the need to ensure digital ID systems do not stifle healthy dissent, which is essential for our communal wellbeing.

The Social Maturity to Dynamically Evolve

Digital ID systems offer significant power but require a level of societal maturity that may not yet exist. These systems risk centralising rigid or dated views, thereby stifling societal evolution by over-simplifying the application of labels. This undermines the dynamic process through which cultures evolve.

History teaches us that challenging norms can lead to advancements. Like the German phrase, *in the bucket*, terms like 'dissenters' and 'criminals' are not substantive arguments. Rather, they are conclusions that have the potential to be misused in defence against challenges to established beliefs. Likewise, endorsing digital ID with simplistic labels like 'secure' or even 'optional' is also problematic. How do we move beyond labels to address these problems?

Reputation: The Language of Trust Beyond Labels

Consider a shift from identity-centric to reputation-centric language for building trust. We have seen how traditionally, trust is expressed in terms of identity, stating: "Paul is making a claim, and the government identifies Paul as trustworthy; hence, the claim is likely true." However, identity merely scratches the surface, offering characteristics, without insight into conduct.

What we truly desire is reputation—a language prioritising actions over attributes. Unlike identity, reputation offers a deeper understanding through behavioural history, reliably verified by shared experiences. It supports anonymous statements like "someone is making a claim, and their previous claims were reliable; hence, the claim is likely true."

The Importance of Reputation in the Real World

Why is reputation the foundation of trust? When facing decisions, we often rely on recommendations rather than unguided research. This preference for personal endorsements highlights our reliance on reputation in various decisions, from selecting products to choosing restaurants. We actively seek reviews because they offer insights into the experiences of others that go beyond basic identifying information, like model numbers, or menus.

Reputation allows for opinions to be evaluated based on personal values, avoiding the interpretative challenges posed by identity, which can vary with timing and context. For instance, a busy executive might favour restaurants with prompt service, dismissing negative reviews about cost as irrelevant. In contrast, a large family may prefer more affordable, leisurely dining experiences. Willing to wait, they value the chance to relax and save money.

Imagine a world where digital systems prioritise reputation over simple identification, incentivising people to act faithfully. This world would foster trust through voluntary interactions, enhancing decision-making to better meet our needs. Imagine no more ...

A Privacy-Centric Reputation-Based Society
Background, Effectiveness, and Motivations

In a reputation-driven society, digital transactions can, and should, occur while sharing as little personal data as possible. Institutional trust is established through verifiable credentials, rather than through sensitive information. This system, which might use the same block-chain technology that protects Bitcoin, can make sure that records of reputation are public, unchangeable, and easy to follow. Such a society is not only plausible but is now more accessible than ever before.

Why should we build such a society? In June 2006, a study titled, *The value of reputation on eBay,* introduced the first randomised controlled trial, exploring online reputation systems. It detailed how an established eBay seller listed identical items under both a recognised profile and new, anonymous profiles he controlled. Findings revealed that buyers paid an 8.1% premium for items associated with the seller's reputable profile, underscoring the online appeal of reputation. (32)

The Importance of Choice in Verification

What do interactions look like without identity? Consider Jane, a school student who must demonstrate to her teacher that she has independently completed her homework. Instead of the traditional method, where she submits her identifiable work for grading directly—subjecting it to the teacher's personal biases—Jane could have the option to select an alternative grader. She might, for example, submit a letter from her parents confirming her work's completion. The teacher could then compare the letter's signature with one on file, bypassing the need for a direct review of Jane's homework.

At the heart of this method is the power of choice. Both Jane and her teacher have the autonomy to choose a mutually trusted grader. If the teacher doubts the impartiality of Jane's parents, due to previous experiences, she has the right to dismiss their endorsement, even if it is genuine.

Jane provides the chosen intermediary with her homework. It is graded and a verifiable credential that represents an assessment of her work is produced. The attestation is uniquely digitally signed, to guarantee authenticity, and then provided to the teacher. This substitution minimises the need for full disclosure and allows the teacher to concentrate on their primary role. Having choice prevents any single authority, like Jane's teacher, from dominating the evaluation of homework.

Outsourcing verification to neutral parties ensures fairness and accuracy, as they have a stake in maintaining their reputation and business. This decentralised approach fosters a competitive environment for trust verification, naturally deterring malpractice. In contrast, centralised systems may actually incentivise compromising integrity.

Additionally, this system capitalises on the expertise of those most qualified for specific tasks. While Jane's parents are well-placed to confirm the completion of her homework, they may lack the expertise to assess its academic quality. A subject matter expert could step in to evaluate the work's accuracy, potentially offering insights superior to those of even the teacher.

Thus, two separate trusted intermediaries could validate both the completion and the correctness of Jane's homework separately. Applying such an approach on a large scale offers a robust alternative to the flaws of trust when based on centralised government digital ID systems.

Trust Agents: A Formal Solution

Let's formalise this idea by introducing the term, 'trust agent'—a new type of intermediary that independently vouches for your trustworthiness in any arena. A trust agent can be anyone or anything that facilitates decentralised institutional trust. Their primary business is reputation, with their success hinging on the accuracy and honesty of their evaluations.

Trust agents prove claims using objective criteria to generate credible and secure digital recommendations, akin to restaurant reviews or parental signatures. They safely decouple the assessment of eligibility to access resources or transact from the personal data used to determine it.

How could trust agents help Jane? As she develops into an entrepreneur seeking business loans, she currently faces the daunting task of sharing vast amounts of personal data with multiple lenders, who each use her identity to determine her eligibility. In a reputation-based society, she no longer needs to expose this data to lenders directly.

Jane could instead employ a trust agent to evaluate her loan eligibility. Their objective determination is then presented to lenders, signed with the agent's unique digital signature to prove its authenticity. The agent's commercial and social reputation is more valuable than any potential bribes from Jane, ensuring they handle sensitive information carefully and maintain referral reliability.

Lenders favour agents with a track record of precise assessments, valuing merit-based referrals over subjective identity judgments. Trust agents leverage personal relationships and proven success stories, while keeping sensitive information, like income or spending habits, hidden from lenders.

Jane secures an institutional referral affirming her creditworthiness, having shared personal details only once. Armed with this, she approaches lenders who verify the endorsement electronically, bypassing the need for each one to scrutinise her entire financial history. The trust agent model revolutionises institutional trust by embedding reputational stakes, motivating all involved to act responsibly. This effectively tackles the problem of misconduct born from an absence of repercussions that are inherent in traditional identity verification systems.

Specialist trust agents bring further meaning to institutional trust. For example, a loan application for a medical business becomes more compelling to lenders when endorsed by a medical trust agent. A referral from an agent staffed by seasoned doctors lends credibility to Jane's application. They evaluate her business plan through the lens of their industry experience, without revealing the plan's details directly to lenders. Jane can secure targeted endorsements from these agents, who access only the necessary information to verify her claims.

Collaborative Trust Agents

Trust agents extend beyond profit-driven companies. They can equally be groups of individuals pursuing a common purpose, perhaps promoting local tourism. Let's introduce the term 'collaborative trust agent' to describe this structure, one that safely migrates relational trust to institutional trust.

How can collaborative trust agents further help Jane with her business venture? Just as Jane can use a financial trust agent to assess her credit status, she can also voluntarily gather digital endorsements from her local community—those who know her best—and use these as character references for lenders. This approach allows for a broader, more personal basis for establishing institutional trust with a lender, leveraging both formal and informal networks to support her credibility.

This collection of reviews from residents acts as an anonymous, yet verifiable petition, providing Jane with the necessary social credibility for a business loan. By contributing to the petition, community members effectively vote in favour of her venture, demonstrating a shared interest in her success. Collaborative trust agents acknowledge the common advantages of Jane's business, such as boosting local tourism, while offering lenders proof of community backing, indicating a higher likelihood of success.

These groups, covering areas from neighbourhoods to entire towns, might involve community members such as the coffee vendor mentioned in the previous chapter. Despite not knowing Jane's name, his long-term observation of her support for his business provides him with an understanding of her character. This unique perspective enables him to effectively endorse her loan application, leveraging his community involvement to assess her reputation, and providing her with the social capital determined so valuable by the eBay study.

Benefits for Providers and Consumers

Under the trust agent model, lenders are motivated to trust verified anonymous referrals, which protect them from internal biases. These recommendations equally eliminate the need for expertise in identity assessment, allowing lenders to concentrate on their core financial services. Lenders benefit from the shared experiences of their peers in accepting referrals from different trust agents. They equally enjoy enhanced public confidence because they do not possess any identifying information about their clients.

Meanwhile, Jane enjoys an unbiased assessment of her eligibility, avoiding the potential conflict of interest where lenders might lower standards to gain business. Her motivation to repay is driven by the need to maintain a good reputation, as trust agents could limit future assessments if she has a poor repayment history.

The trust agent idea represents just one of several possible approaches to building a healthy society that does not rely on identity for every interaction. While no system is foolproof, those that incentivise idealistic behaviour, encourage the majority to act in good faith by reducing the appeal of corruption. With their ability to operate despite bad actors, trust agents present a strong argument for shifting resources from government digital ID to more innovative and effective alternatives.

Implementing Change

Strategic Rollout and Adoption

Could we really eliminate the use of identity when building institutional trust? Exploring how these ideas can be realistically implemented makes the concepts feel practical and within reach. Historical examples, including the Enlightenment, the Industrial Revolution, and the rise of the internet and social media, illustrate humanity's capacity for significant technological and societal change. These precedents support the feasibility of reducing government influence over trust, potentially marking a shift as pivotal as the separation of church and state.

Phased and strategic deployment is key to successfully implement a reputation management system. This allows for continuous improvement based on user feedback. Collaboration with experts in technology and cybersecurity can help ensure the system's effectiveness, making it reliable and commercially sound.

Educating users about the system's benefits can encourage widespread adoption, helping them understand and embrace its features. Equally, implementing safeguards against reputational sabotage is essential for ensuring the system's resilience.

Resilience to Reputational Sabotage

Reputation can be quickly lost. Can we clearly differentiate between genuine reputational loss and that caused by malice? To safeguard against reputational sabotage, reputation systems should incorporate granularity. This allows for evaluating distinct facets of a person's behaviour, offering a nuanced perspective that is more challenging to manipulate.

Accumulating historical data over time provides a comprehensive view of behaviour, making significant reputation changes without genuine behaviour modification challenging. Permanent records help spotlight any anomalies, while decentralisation distributes these records across multiple entities. This prevents any single actor, like a government, from having enough control to alter the reputation landscape significantly. Any manipulation attempts become more visible, potentially even damaging the perpetrators' reputation.

Transparency in the calculation of reputation scores, and the provision of appeal mechanisms, ensure fairness and accountability, allowing people to contest negative ratings.

There are many ways to mitigate the risk of unfair manipulation in fostering a trustworthy society.

Dispute Handling

What benefit is there in winning disputes if it results in creating future adversaries? In a reputation-based society, disputes become calls for help, rather than adversarial exchanges. Moving on from identity refocuses dispute resolution on broader systemic reforms, encouraging people to consider the collective good beyond just resolving their own issues.

Embracing trial and error, and community feedback, is key to improving dispute resolution. Fairness is ensured through transparent resolution processes, employing methods like voluntary arbitration, peer review panels, and reputation audits to resolve conflicts effectively. A public ledger that records case outcomes can help establish clear standards. Moreover, providing Conflict Resolution Training to the community equips people with vital negotiation skills, reducing dependence on formal dispute resolution mechanisms.

While the exact solutions in a truly free market remain uncertain, numerous promising ideas have been proposed. For instance, philosopher Stefan Molyneux highlights the potential role of what he calls Dispute Resolution Organisations (DROs) in stateless societies. He defines them as:

Companies that specialise in insuring contracts between individuals, and resolving any disputes that might arise. (33 p. 57)

By blending existing modern technologies with conventional trust and expert opinion, this society offers high-quality dispute resolution that challenges traditional adversarial models.[48]

Integration with Existing Frameworks

Parallel implementation with current systems is key to introducing a reputation-based society. Recent examples, like the emergence of ride-sharing services within the traditional taxi industry, demonstrate that reputation-based trust, facilitated by driver and rider ratings, can effectively coexist alongside traditional identity-based systems reliant on government licensing and identification. This

---

[48] Such modern technology may include smart contracts, popularised by the Ethereum block-chain.

provides users with options and allows market dynamics to favour the most effective solutions.

Implementation will be gradual, initially complementing existing legal and social frameworks. For instance, reputation systems could offer deeper insights into past behaviours, enhancing traditional credit evaluations. Thoughtfully integrating reputation with current systems can lead to a society less dependent on traditional enforcement.

Novel ideas often lack foolproof roadmaps. Yet clinging to flawed, harmful systems for lack of a fully fleshed-out alternative hinders progress. Once we commit to change, we can explore and refine new solutions. Let's make choices today that free us from government digital ID, securing a legacy we can be proud of.

Chapter Takeaways

In the spirit of fostering a society that values trust and reputation over imposed digital ID, this section offers actionable advice to help readers protect their identities while cultivating genuine integrity that improves their lives. These takeaways aim to empower people to make choices that align with the vision of a more trust-oriented future.

In Private

Embrace Privacy as a Positive Value

Challenge the misconception that privacy implies secrecy or guilt by recognising it as a fundamental right. Advocate for a positive view of privacy to counter the flawed idea that "nothing to hide means nothing to fear." Promote discussions about privacy's role in building trust and a solid foundation for a reputation of integrity over time.

In Public

Strengthen Communities Through Integrity

Cultivate a reputation for integrity across all areas of your life, from personal to professional. A solid reputation acts as social credit, enabling trust-based interactions without invasive identity checks, just as with my tool borrowing neighbour. Strengthen community bonds through consistent, honest behaviour and support local initiatives like neighbourhood watch programs. By strengthening relational trust, you contribute to a mentality that values actions over labels and prevents governments from muscling in to upset the balance.

Practice kindness and allow for both disappointments and positive surprises. While you might risk losing a tool loaning it to your neighbour, you could gain a supportive friend. Surrounding yourself with people who share common values makes life's challenges, including government overreach, more manageable.

Enhance Communication with Precise Language

Socrates famously stated that the beginning of wisdom is the definition of terms. Misinterpretations often stem from assuming a mutual understanding of words such as 'convenience', 'trust', and 'criminals', which can cause disputes, particularly in policy debates. Committing to precise language can prevent many conflicts and is applicable in both personal and professional contexts.

Encourage clarification when someone uses a word, even one that seems straightforward. Asking for their definition isn't rude; it shows a genuine interest in their perspective and a commitment to better communication. This approach cultivates a culture of mutual understanding and respect.

The use of simplistic language often stems from a habit of deferring complex issues to bureaucrats. Yet, for societal progress, each generation must courageously tackle cultural challenges themselves. Being labelled as dissenting in doing so can be empowering, as it indicates we are providing value through diverse opinions. Ideas that lack merit will fail naturally under scrutiny, while credible ones will stand on merit and enhance our future. By critically assessing the terms used by governments to justify digital ID, we walk the path to wisdom that Socrates outlined.

Cultivate Trustworthy Societies by Being Selective

Assess where you may be misplacing trust in people or organisations. Trust must be earned through consistent actions, not just assumed. Withdraw support from individuals and institutions who fail to keep their promises and direct your resources towards those that align with your values and demonstrate integrity.

Our aim is to foster societies rich in trust, where strong, genuine connections reduce the need for constant monitoring. By thoughtfully engaging with high-quality relationships and distancing from those that do not meet our standards, we uphold our dignity.

Doing so simultaneously offers constructive feedback to those falling short, encouraging them to elevate their standards and reintegrate into our circles with improved practices. This self-regulatory approach diminishes reliance on pervasive identity systems, cultivating a culture where accountability is naturally embedded.

Finding Happiness in Building Trust and Reputation

This chapter emphasised trust as a fundamental element of society and discussed the dangers centralised digital ID poses to its integrity. We explored the variable nature of trust, the dangers of misuse by authorities, and the misleading application of labels to justify control. Advocating for a reputation-based culture with decentralised technology, we presented a privacy-centric alternative to identification. While detailed implementation remains to be fully addressed, prioritising earned trust over imposed identification suggests a promising path forward.

Yet, as we draw the curtains on this chapter, we must recognise the hope that thinking outside of the identity box offers. Trust is more than a social construct; it is a form of currency that enriches personal fulfillment. Fostering a positive reputation promotes self-improvement and strengthens values that build community. By prioritising integrity, we not only enhance our own lives but also unlock a unique source of happiness—contribution.

Trust is inherently reciprocal. In a reputation-based society, we teach our children the value of integrity, a stark contrast to the coercive nature of government digital ID. This shift has transformative intergenerational implications, highlighting the difference between identity-based and reputation-based living. To understand these two futures better, let's now transition from the infinite nuance of trust to the binary precision of technology.

CHAPTER 4:
Technology

Your scientists were so preoccupied with whether or not they could, they didn't stop to think if they should.

—Dr Ian Malcolm, fictional mathematician, *Jurassic Park*, 1993 film. (34)

In the past, dictatorships have always come with hob-nailed boots and tanks and machine guns, but a dictatorship of dossiers, a dictatorship of databanks can be just as repressive, just as chilling and just as debilitating.

—Professor Arthur Miller, contributing to the Joint Select Committee on the Australia Card, 1986. (12 p. 204)

Mankind barely noticed when the concept of massively organised information quietly emerged to become a means of social control, a weapon of war, and a roadmap for group destruction.

—Edwin Black, *IBM and the Holocaust*, written 2001. (35 p. 7)

Information technology is a new and powerful technology which is still developing rapidly and providing techniques for the exercise of power which will be ill-understood for some time.

—Graham Greenleaf, University of New South Wales Law Lecturer, and New South Wales Privacy Committee member. (36)

Erwin Cuntz: The Father of Tech-Driven Identification

In 1934, shortly after the Nazis assumed power, Freiburg lawyer Erwin Cuntz sat down to write a personal letter to Hitler. That letter changed the world. In it, Cuntz advocated for identification policies based on a new concept he termed the Personage Principle.[49] He argued this was essential for aligning human affairs with technological progress. (4 p. 35) This led the Nazis to integrate technology with identity early in Hitler's reign, establishing Cuntz as the father of modern, technology-driven identity management.

This chapter reveals the significance of that letter, by exploring technology's impact on human identification and breaking down complex engineering concepts for clarity. Understanding how the Nazis used technology to collect and process data provides invaluable insights into digital ID, including its appropriateness and implications. We will discuss the nature of digital data, its potential for reinterpretation, and the critical need to sceptically evaluate new technologies and their centralised control. Highlighting the ethical concerns and lasting societal impacts of digital ID, we explore the importance of vigilance against misuse, while acknowledging technology's benefits ... but first, back to the letter.

The Personage Principle and the *Volkskartei*

Born in 1872, Cuntz had long recognised the authoritarian potential of centralising identity records within a single bureaucracy. He sought to use emerging technologies to vastly increase the collection and use of personal data on a scale beyond regional limits, envisioning a government with unprecedented knowledge of its people. His innovative proposals are easily missed, yet they form the basis of modern centralised identity registries.

---

[49] *Personalitätsprinzip.*

Restructuring Identity Data

Historically, population data was organised by variable qualities, like residential addresses, rather than fixed qualities such as ethnicity and birth date. This served statistical and planning needs well. Yet, finding someone by name was not a primary function, making it a cumbersome, needle-in-the-haystack task.

The Personage Principle marked a significant shift in data management, moving from a geographic focus to an individual focus. The prioritisation of identity attributes, discussed in Chapter Two, was key to developing the Nazi population registry, the *Volkskartei*. It allowed for rapid government access to personal records, bypassing the protective inefficiencies of searching through household-based data to find specific people.

To maximise the *Volkskartei's* utility, Cuntz emphasised the need for full participation and proposed strict penalties for non-compliance. Hitler, recognising the letter's potential for total control, quickly approved the plan, and the *Volkskartei* went into service in February 1939.

The Volkskartei in Action

With the help of schools, registration in Germany was initially mandated for all aged from 5 to 70. (37 p. 130) Caught in this broad age range was my then teenage *Oma*. This prompted a strong reaction from her father, Fritz, wary of the rising domestic militarisation of the previous years.

Fritz viewed these developments, enforced with penalties, as violations of his daughter's innocence, triggering a deeply protective paternal instinct. Cuntz had proposed forcing families to report residence changes, extended travel, employment shifts, births, and deaths. (4 p. 35) The system also linked appraisals of mental and physical fitness captured by the 1925 "economic and social-statistical evaluation" census.[50] (4 p. 16) Fritz felt the government had claimed a seat at the family table, passing judgement on private affairs.

Yet despite his rank and rage, there was little Fritz could do to shield his family from the harsh penalties for non-compliance that Cuntz had advocated. The system was already operational, allowing the regime's elite to conduct immediate, targeted identity searches, enforced by police authority.

Like many such systems, the *Volkskartei* expanded to include details like religion, and military-relevant skills including higher education and driving licences.[51, 52] Later, it grew even more intrusive, further capturing physical traits, including head shape, nose size, eye placement, and cheekbone structure.[53] (4 p. 15)

The *Volkskartei* replaced decentralised, regional registrations with a uniform, centralised system, removing previous protections. It required applicants to complete gender-specific application forms: women reported skills in office tasks and domestic work, while men provided vocational details, travel history, education, and experience in fields like agriculture and science (Figure 10). (38 p. 126)

---

[50] The data captured in this 1925 evaluation strikingly resembles some questions from the 2021 Australian Household Census. These include Q24: *"Does the person ever need someone to help with, or be with them for, self-care activities?"* Q25: *"Does the person ever need someone to help with, or be with them for, body movement activities?"* Q26: *"Does the person ever need someone to help with, or be with them for, communication activities?"* Q27: *"What are the reasons for the need for assistance or supervisions shown in Questions 24, 25 and 26?"* Q28: *"Has the person been told by a doctor or nurse that they have any of these long-term health conditions?"* (46 p. 10)

[51] Similar to Q23 from the 2021 Australian Household Census, *"What is the person's religion?"* (46 p. 10)

[52] Similar to some questions from the 2021 Australian Household Census. These include Q29: *"Is the person attending a school or other education institution?"* Q30: *"What type of education institution is the person attending?"* Q32: *"What is the highest year of primary or secondary school the person has completed?"* Q33: *"Has the person completed any educational qualification?"* Q34: *"What is the level of the highest qualification the person has completed?"* and Q35: *"What is the main field of study for the person's highest qualification completed?"* (46 pp. 12-14)

[53] It is hard not to draw parallels between these records and today's facial recognition technology.

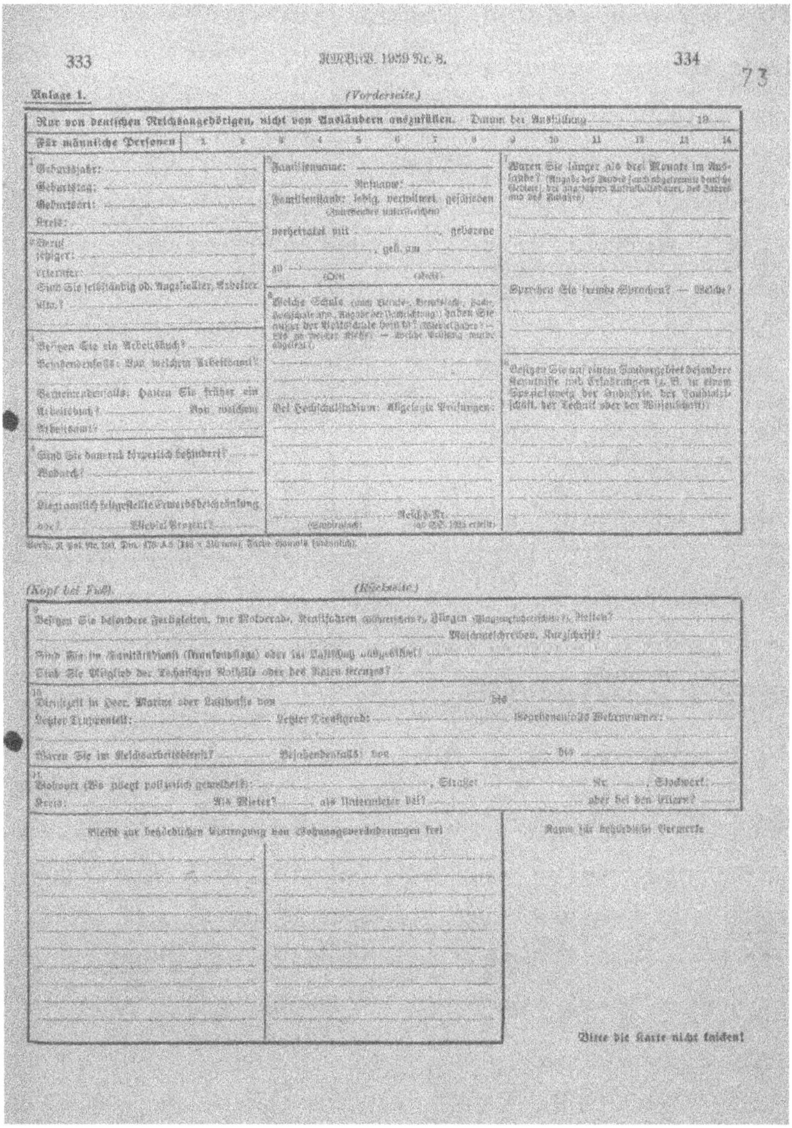

Figure 10. *Volkskartei form for men; front and back (39 p. 148)*

The Beginnings of Computer Design
Data Storage and Access

How best to organise this information? The physical design for the *Volkskartei* was practical and elegant. Cuntz proposed a 25-story cylindrical tower, with each floor devoted to birth date intervals offset by 25 years. The floors were divided into twelve wedge-shaped rooms for each month, containing up to 31 cabinets corresponding to each day of the month. This layout allowed quick access to records, sorted by birth date.

The tower's design closely mirrors that of magnetic hard-drives, essential parts of modern digital data storage. These consist of circular platters mounted on a single spindle, like the tower's floors. In *The Nazi Census,* Aly and Roth note that each hard-drive platter is further divided into tracks and sectors where data is written, reflecting the tower's organisation into rooms and cabinets. Both designs maximise data density and allow efficient retrieval.

Yet the comparison extends past Aly and Roth's insights. Organising the tower's floors in 25-year intervals balanced the distribution of cabinet weights. This concept mirrors modern so-called 'write balancing' algorithms, software that evenly distributes data across a digital hard drive to prolong its lifespan.[54] Such foresight by Nazi innovators in managing data efficiently, laid the groundwork for sustainable engineering designs still used today.

Equally, the tower's method of sorting information—initially by birth date and then by other details—mirrors modern data organisation in computer systems. Today, we categorise identity data by attributes such as gender, name, and birthplace, using computer commands

---

[54] Write balancing is especially pertinent for solid-state drives (SSDs) and flash memory, which have a finite number of write cycles. In magnetic hard drives, similar strategies are employed to prevent undue wear on specific disk areas. Technologies like RAID (Redundant Array of Independent Disks) also distribute data across several disks to balance the load and minimise wear on single drives while providing redundancy.

to sequence it efficiently.[55] This approach, consistent through time, underscores a continuous goal: to simplify the management of otherwise unwieldy masses of population data. It demonstrates the enduring impact of technology on our identities.

Volkskartei Cards

The registry cards in the filing cabinets also showcased early innovations in data management. Each card had numbers at the top edge representing categories such as religion, with colour-coded tabs attached like Post-it notes to indicate specific values. For example, a red tab next to the religion number identified the individual's specific faith.

The protruding tabs functioned as bookmarks, enabling quick identification of specific traits without disturbing the stacked cards. This streamlined searches and enhanced storage efficiency. Tab numbering was standardised; the left half was designated for local identifiers, such as volunteer fire brigade members and vehicle owners, while the right half was reserved for national purposes. (38)

The tab system also featured a new mechanical querying aid: officers threaded a wire through aligned holes in tabs representing specific traits, such as occupation. By lifting the wire, they could swiftly extract matching cards, minimising human error, and eliminating manual searches (Figure 11).

Lessons Learned from the *Volkskartei*

The *Volkskartei*, a precursor to digital ID, set the stage for the Nazi regime's surveillance technology. It forcibly gathered detailed personal data, providing the government with rapid intelligence for targeted oppression. More than a technology milestone, the *Volkskartei* underscores the dangers of centralised data, and the seductive

---

[55]   One parallel is the modern SQL 'ORDER BY' command, which sorts data by multiple criteria.

Figure 11. *Nazi officers had unrestricted access to highly granular population identity data provided by the Volkskartei identity registration system. Storage and searchability of the centralised identity cards proved innovative and greatly increased the speed and scope of damage that the Nazi Government was able to inflict on the people of Europe. (4 p. 3)*

allure of such invasive systems. Its legacy stresses the imperative of scrutinising the intentions behind, and possible misuses of new technology before its adoption.

The *Volkskartei* did not simply materialise overnight. Instead, it relied on targeted and persistent propaganda that highlighted supposed benefits, and stoked fear to ensure its acceptance. Its success was largely due to the public's lack of understanding regarding the power of data, and the resulting power imbalance it creates between the government and the individual. This oversight allowed hidden agendas to drive its adoption without significant opposition.

Technology's Dual Nature: Saviour and Executioner

As an engineer, I respect technology's power to elevate living standards and life expectancy, but I'm also wary of its flip side. There is a darker side to technology. It can invite everything from job displacement via industrial automation to destruction via military applications. Safely integrating technology into society requires acknowledging its dual potential for both good and ill.

This section provides that balance. The Nazi era showcased the dangers of technological advancements under malevolent control, described by author Byron Richards as a time when:

... the powers of science were usurped by a madman. (15 p. 54)

To understand how this played out in the context of identity management, let's examine a specific innovation developed for Hitler's sinister goals by a company called *Dehomag*.

Dehomag: The Birth of Computerised Identity Management

In the early 20th century, a technological leap emerged from Berlin, forever changing the identity management landscape. Founded in 1910, *Dehomag*, a subsidiary of the company later known as International Business Machines (IBM), arguably birthed the computerisation of human identity.[56] (4 p. 11). Its initial mission was to enhance administrative efficiency, yet *Dehomag's* contributions provided the Nazis with the tools for rapid and systematic persecution of millions, based on detailed identity records.

The Punch Cards: Categorising Individuals

At the heart of *Dehomag's* technology were machine-readable punch cards, known as *Lochkarten*. These cards encoded personal details like nationality, birth date, and marital status into a standardised row and column format for automated processing, known at the time as 'mechanisation'. This system distilled individuals into

---

[56] *Dehomag* is as an abbreviation for the *Deutsche Hollerith Maschinen Gesellschaft* (German Hollerith Machine Company). The company was named after German-American engineer, Hermann Hollerith, who was employed by the US Census Bureau at the end of the nineteenth century. (4 p. 10)

quantifiable attributes, represented by holes punched in specific card locations.

The use of punch cards to categorise people was transformative. The unsettling way it blended technology with our human essence was captured in a *Dehomag* procedure document of the time. It espoused the benefits of mechanising the core tasks in 'accounting'—sorting, writing, and arithmetic—claiming:

> *... the means to achieve this purpose are the punch card and the machines necessary to process it. (40 p. 24)*

The Machines: Precise Data Entry and Reporting

Dehomag's machinery, precursors to digital databases, included models 20 104 and 20 108 which used a numerical keypad to punch data onto cards (Figure 12). Data-entry clerks encoded population information by punching holes in designated rows using a magnetic mechanism triggered by keystrokes (Figure 13). Additionally, *Dehomag's* hole tester models 21 004 and 21 008 enhanced error checking. These units used feeler pins to detect discrepancies between keyed values and actual punched holes, marking errors on the card's rear to ensure data accuracy before processing.

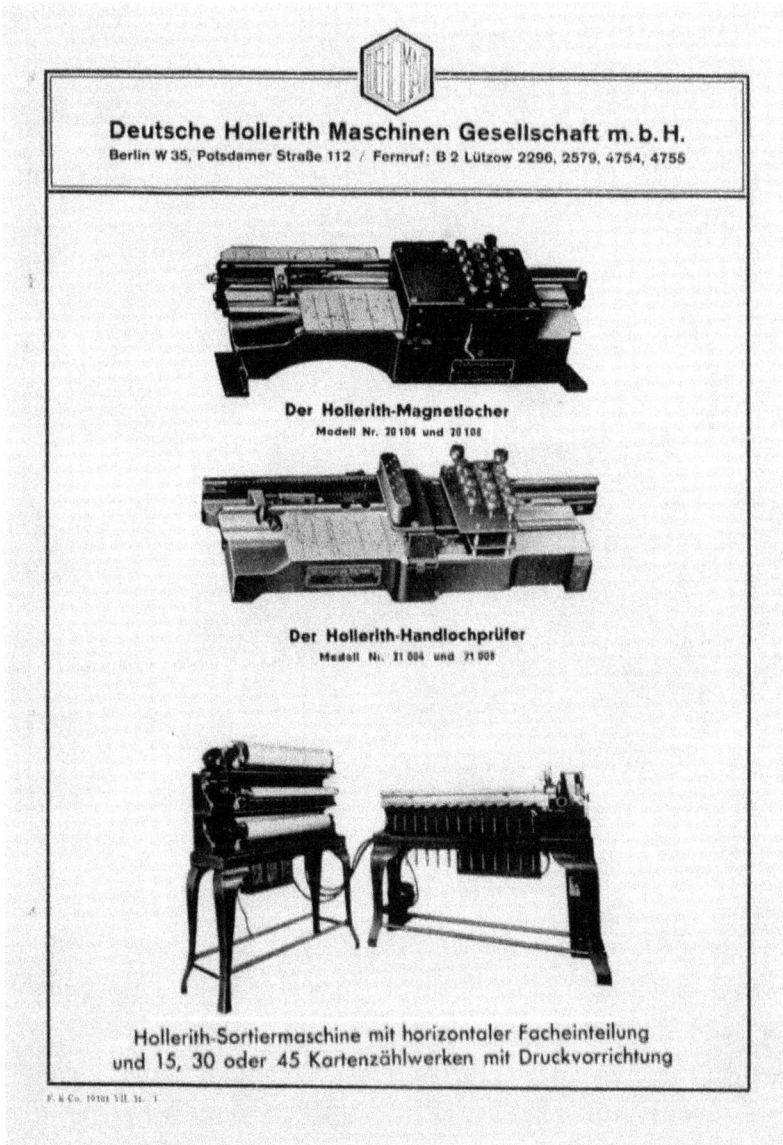

Figure 12. *Combined images of the Dehomag magnetic card punch machine (40 p. 19), hand card hole tester (40 p. 20), and card sorting machine with horizontal compartment division and printing device. (40 p. 27)*

Figure 13. *Clerks encoding population data onto Dehomag punch cards. An instructional sign adorns the wall. (41)*

The sorting machines, operating on 110 volts DC, processed up to 24,000 cards per hour—effectively analysing about seven people each and every second. Equipped with automatic card-overfill protection, these machines efficiently and accurately sorted identity cards based on the data encoded in specific columns, organising the results in ascending order and generating summary reports. This represented early integration of data processing and reporting.

Dehomag machines automated the processing of identity records by writing, reading, and sorting cards based on specific criteria. This allowed Nazi officers to quickly locate people with particular characteristics, such as married Catholic parents of a certain age in a specific region, or French speakers with a history of vagrancy arrests. The widespread use of *Dehomag's* equipment necessitated regular maintenance by company staff to maintain continuous operation, ensuring that no one was beyond the system's reach. (35 p. 22)

Automated Discrimination: A Profitable Business

Dehomag's collaboration with the Nazi government was deeply complicit. In a letter to the Reich Insurance Office dated February 17, 1934, *Dehomag* expressed its eagerness to support the regime's objectives:

We have again carefully calculated the work processes discussed with you and would like to agree to carry out this work for you. (42 p. 80)

Dehomag incentivised this business by offering the government installation services, a five percent discount on machine rentals, and favourable terms for fixed contractual relationships:

… in order to continue this work or for other purposes. (42 p. 80)

Their monopoly positioned *Dehomag* as the sole supplier of the 1.5 billion punch cards needed annually for the Nazi extermination program. (35 p. 22) They were also tasked with transferring the 1933 German census data onto punch cards, a project prioritised by Hitler upon taking power. This work was performed at the Reich Department of Statistics and was swiftly used to identify and strip citizenship from those born abroad.

Unsurprisingly, this mechanised punishment system reached beyond those of foreign birth. Enacted on July 14, 1933, the *Law of Revocation of Naturalisation and the Revocation of German Citizenship* legally stripped citizenship from those simply deemed 'undesirable'. (4 pp. 59-60) With *Dehomag* technology, machine operators could easily brand people with this label by adjusting the search settings, producing a sorted stack of identity cards, detached from the people they represented. Technology let the government delegate power to staff who made decisions based on personal biases, even down to the purity of my *Oma's* German blood.

Private Business and Forced Labour in the Holocaust

By 1942, *Dehomag* had even secured as a client the infamous Nazi Protection Squad, known as the *Schutzstaffel* or *SS* (4 p. 14). The SS Race Office used this technology to survey and evaluate people, classifying the results (Figure 14). For instance, a '6' punched

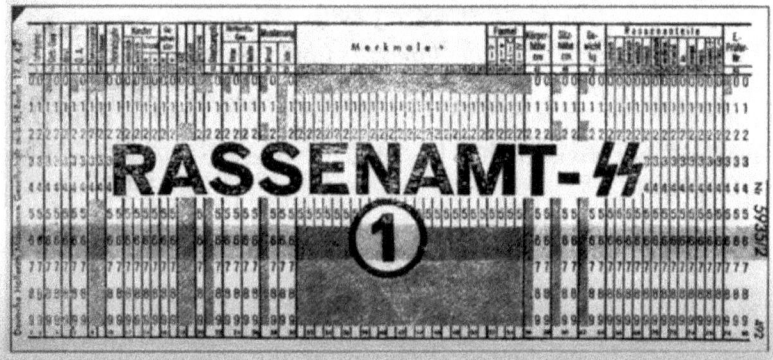

Figure 14. A Nazi SS Race Office punch card with numbers in different columns. (4 p. 15)

in column 34, labelled 'Reason for Departure', indicated *Sonderbehandlung* [special treatment], a coded term for execution. (35 p. 21) The government could easily impose identity-based punitive actions on people once:

... numbers and punch cards had de-humanised them. (35 p. 22)

As new prisoners arrived, their corresponding punch cards were fed into *Dehomag* machines. This system efficiently allocated prison labour to local farms and factories requiring specific language skills or qualifications. Like shuffling a deck of playing cards, the machines selected people with a mechanical whir and buzz, printing summary manifest sheets that followed those chosen to their machine-designated destinies. This process occurred entirely without human touch or moral judgment, treating people as mere data points.

The technology sector's role in the Holocaust was profound, actively enabling genocide via equipment, training, and maintenance. *Dehomag* technicians servicing machines at death camps were directly exposed to the horrors, literally inhaling the stench of burnt "column 34, code 6" flesh spewing from the crematoria as they worked. They proceeded anyway.

The broader German private industry also benefited from these practices via the use of targeted slave labour. Even before the war, the *SS* had established partnerships with private firms to exploit government-facilitated slavery under Heinrich Himmler's supervision. *Auschwitz*, the camp where little Nadine died, was a prime example of this dark and self-sustaining practice.

Dehomag technology, in tandem with government policies, generated a vast supply of slave labour. The lure of war profiteering saw this ruthlessly exploited for mutual gain. Albert Speer, becoming Minister of Armaments and War Production on February 8, 1942, used this workforce to strengthen the military. In March, Hitler mandated the conscription of slaves from occupied territories to free Germans to fight. (3 p. 534) This synergy of technology and policy intensified suffering, while rewarding the likes of *Steyr-Daimler-Puch* and *IG Farben*.

Lessons Learned from Dehomag

What lessons can we draw from historical uses of technology in identity management? Clearly, technology has proven highly efficient at forcibly extracting the valuable human resources discussed in Chapter Two. In Germany, *Dehomag* equipment became synonymous with 'labour assignment', with most camps establishing a

Hollerith Department, named after the company's founder, within the Labour Assignment Office.

Using identity to unlock the lives of others was so lucrative that the *Dachau* camp even shielded its *Dehomag* installation within a large concrete bunker, ensuring operation during Allied bombings. (35 p. 428) Internal Nazi communications emphasised the critical need to keep prisoners and their corresponding identity cards together, noting that separation disrupted the system's efficiency. (35 p. 430) This was technology's dual nature on display, along with the involvement of engineers and technicians in facilitating these crimes.

In days gone by, one could be liberated from these creations via errors and external events like bombings. No such liberation from digital ID is possible. Modern biometric systems are inseparable from our physical bodies and can continuously monitor us with cameras in unavoidable places. This raises a crucial question: can modern technology prevent the abuses of the past?

The Protections of New Technology

Have lessons learned from historical abuses of power led to the creation of digital ID systems equipped with safeguards to prevent similar abuses? Supporters suggest that the current implementation of digital ID occurs in a context more attuned to security concerns, and bolstered by stronger legal protections, making past abuses less relevant.

Yet technical security measures only work when those in power implement them. They are ineffective when lawmakers lack technical expertise and ignore input from experts. This was shown by Australia's recent efforts to introduce a centralised government digital ID. The public response to the Trusted Digital Identity Framework (TDIF), included many dissenting submissions, notably one from

Associate Professor Vanessa Teague, who criticised the scheme for its failure to adhere to protections offered by modern technology: [57, 58]

> *Neither the TDIF's high-level design, nor its implementation by the ATO (myGovID) meet their intended security goals.* [59]

> *The myGovID system is subject to an easily-implemented code proxying attack, which allows a malicious website to proxy a person's myGovID login and re-use their authentication to log in to the victim's account on any website of their choice.*

> *Although detectable by extremely diligent users, the attack is likely to go unnoticed by most victims. A video demonstration is available at https://youtu.be/TgPdVbUbtBM.*

> *This was disclosed to the ATO in August 2020, but they informed us that they do not intend to fix it.*

> *The system should be abandoned and redesigned from scratch by people with some understanding of secure protocol design and some concern for protecting their fellow citizens from identity theft.*

> *Legislating to make it secure by fiat will not stop organised crime, foreign governments, or ordinary criminals, from taking advantages of its design flaws. (43)*

Good intentions do not guarantee ethical outcomes. Safeguards may fail or simply not be implemented in the first place. Despite technological advancements, the core issue of power imbalance persists. Do new technologies worsen rather than mitigate this issue? As we reflect on *Dehomag's* analogue technology, it becomes imperative to understand how the evolution of digital systems has transformed our ability to record and process human identity.

---

[57] The Federal Government started development of the Trusted Digital Identity Framework (TDIF) in 2016, launching it in early 2018.

[58] CEO, Thinking Cybersecurity Pty Ltd. A/Prof (Adj.), ANU.

[59] Australian Taxation Office (ATO).

Analogue and Digital Systems

Data is the modern fuel that flows seamlessly through everything, from vehicles to consumer devices. Yet, despite its ubiquity, the essence of data often goes unexamined. Think of data as measurements about interesting things. This may include recordings of orchestral music, annual rainfall, election votes, or identity traits. The way we handle these measurements influences our relationship with them. A basic grasp of data recording and processing is crucial for a meaningful discussion on digital ID; and is the focus of this section.

The Analogue Domain: Real World Complexity

Much like our identities, real-world phenomena resist simple categorisation. Instead, they flow continuously, like the pressure waves interpreted by our ears as music. Capturing data about our infinitely complex world proves challenging due to this inherent variability.

In the pre-digital age, recording data required systems of equal variability. These directly encoded the wide range of possible real-world measurements onto a physical medium, like the grooves in vinyl records. Engineers term these measurements 'continuous', and the equipment handling them 'analogue'.

Eight decades ago, the Nazis showcased the possibilities of the era's limited analogue technology. The *Volkskartei* identity cards, for example, were tangible embodiments of life's variability. Like fingerprints, each card bore unique handwriting, ink density and paper texture. This posed challenges for mass production, legibility, processing speed, and consistency.

Dehomag machines were tactile, analogue creations with moving parts requiring lubrication. They demanded careful handling to avoid jamming when processing cards, with technicians relying on audible cues from mechanical parts to troubleshoot issues. The machines sorted identity records just like a human would, albeit with higher speed and fewer errors. Yet, the constraints of the analogue domain spurred inventors to seek alternatives.

The Digital Domain: Simplistic Detachment

Digital technology brought a new approach to recording data, diverging from directly replicating the world's variability. It introduced the concept of 'sampling', which estimates values at specific time intervals. For example, recording music digitally involves measuring the sound wave pressures thousands of times a second, storing these independent values to be pieced together later. Processing these samples, rather than the raw data itself, helps tame the complexity of analogue life. Engineers term these measurements 'discrete', and the equipment handling them 'digital'.

Compared to analogue counterparts, digital systems offer convenience in encoding and processing. The key lies in imperceptibly and flawlessly reproducing the original source by storing samples in a way resilient to the challenges faced by analogue data. Even if some samples are corrupted, redundancy ensures usability.

However, digital systems have a curious and disturbing property—they detach us from our data. In the digital realm, we lose the sensory connection to our world, as we can neither hear samples nor see the binary code representing them. Instead, we rely on software, and by extension the engineers who write it, to interpret the data on our behalf.

Think of a wristwatch to grasp this distinction. An analogue watch, its hands moving continuously around the clockface, can represent an infinite number of time points, leading to subjective and unique interpretations like 'nearly three o'clock' or 'about five to three'.[60] This continuous variability makes analogue data difficult to store,

---

[60] The same concept is obvious with the use of sundials.

categorise, and process. On the other hand, a digital watch simplifies timekeeping by displaying time in finite, precise intervals, such as 02:56. The limited number of values reduces the effort of interpretation.

Key Properties of Digital Systems

Some properties of digital systems make their application to identity management worrisome. Their ability to represent data as samples allows for efficient storage, sorting, copying, and transformation at great scale and speed. These are some of the powers of digitisation.

Durability

Digital data stands out for its ease of replication without quality loss, unlike the degradation seen in copying analogue formats like vinyl records. Advanced error correction ensures digital integrity over time, while analogue copies deteriorate. Affordable, robust hardware allows for cost-effective data backups, safeguarding against loss. Consequently, digital systems are durable against wear and decay, enabling the perpetual preservation of pristine data.

Availability

Digital data is hard to delete. Modern communications have transformed data availability, providing high-quality, remote connections that far surpass the limitations of crackly, tethered analogue equipment. However, this advancement comes as a double-edged sword. While digital technology has enhanced accessibility, it has also complicated data permanence. Even intentional erasure can leave residual traces, necessitating advanced tools to ensure complete removal.

This challenge was highlighted by the 2016 Hillary Clinton BleachBit scandal. BleachBit is a software designed to thoroughly delete files. It was used to wipe Clinton's personal server, after a subpoena was issued by the House Select Committee on Benghazi, demanding the turnover of digital correspondence related to the 2012 attacks in Libya. Erasing data from a single computer is one thing, but once shared, such as through an email, digital data becomes vulnerable to the actions of others in an environment where tracking its custody is impossible.

Editability

Does sampling accurately capture the world around us? Modern software allows for simple alterations of digital images, video, and audio, enabling everything from meme generation to falsely implicating people in crimes. This potential for data manipulation, especially for harmful purposes, stands in contrast to analogue mediums like film photos or paper documents, which are less prone to tampering and often provide a truer representation of reality, imperfections included.

Speed

Lastly, digital systems are fast. Data samples transcend time, seen in the immediate time-setting of a digital watch versus the manual winding of an analogue one. Their independence allows for instant, non-linear access to information, such as selecting a specific scene on a streaming service. This stands in contrast to the slow, linear search of analogue methods like fast-forwarding videotape, or even climbing the 25-storey Cuntz tower.

Rethinking Digital Dominance

Is digital technology inherently superior? Or is this simply a misconception? Despite these remarkable properties, the best technology for any application depends on the context. For example, digital watches, without moving parts, are durable and well-suited for military use, while analogue watches can be adjusted without complex instructions, and are favoured in hospitals, schools, and clock towers for their readability. Additionally, the aesthetic appeal of analogue watches, as seen in the luxury brands favoured by James Bond, highlights their popularity as an investment among society's affluent.

When engineers make design decisions, they weigh various trade-offs. Embracing the benefits of digital systems also means accepting their drawbacks. This choice should align with the specific needs of the application, challenging the idea that digital solutions are always superior. Navigating these trade-offs reveals that the challenge is not just about choosing technology, but critically evaluating the nature of data itself. Doing so reveals how subjective our interactions with digital information are.

Data versus Data Evaluation

Interpreting Digital Identities

How is data interpreted? How do the properties of digital systems influence this interpretation? When we talk about digital ID, we are essentially asking the fundamental question: Can the accuracy of digital data ensure an equally accurate understanding of our world?

On February 28, 2024, American political commentator, Tucker Carlson, shared insights from his recent trip to Moscow, where he interviewed President Putin. Carlson made the following astute observation about how information is interpreted:

The digital information sources that I used to understand just something as simple as 'what's the city of Mocow like' were completely inadequate. (44)

Carlson's remarks find relevance far beyond the confines of Russian cities. The interpretation of many everyday experiences, such as the beauty of a sunset, varies by context and is inherently subjective.

Questions like, "How beautiful is this sunset?" lack definitive answers because beauty cannot be quantified easily. While we can record objective data—such as cloud coverage, colour, and duration—the subjective nature of our assessments, influenced by time and the observers' moods, remains. This section explores how digital ID, like a sunset, is interpreted differently by different people and can change over time.

A Legal Example

Does the legal system offer a perspective on the difference between data and its interpretation? In court, evidence is data, and the jury's verdict is the interpretation. Think of this like the difference between a sunset's measurements, and an evaluation of their beauty. Recording verdicts without preserving evidence prevents future re-examination with new technologies like DNA analysis. This deprives future generations of the chance to reassess historical cases with a modern understanding. The outcome? We potentially leave a tainted time capsule representing the analysis of people of the period.

Preserving raw evidence can be important. It opens the possibility of re-trials and re-evaluation of past events, correcting errors and potentially creating new ones, by evaluating historical events against current standards. This concept is so important that the preservation of source material in journalism was one of Julian Assange's motivations for the formation of Wikileaks. (45) However, considering the subjective nature of data interpretation, we must recognise that changing policies can lead to the retroactive reinterpretation of data, impacting individuals significantly—a lesson underscored by history, and relevant to today's digital ID discussions.

An Engineering Example

Over my career, I have often faced the distinction between raw and processed data. In one instance, I was diagnosing a fault in some navigation software. Under certain unknown circumstances, the error had produced bad directions for a customer and, as an engineer, the first task was to examine the software logs for clues to its cause.

The software was supposed to track the movement of a vehicle by recording its coordinates—the specific points on the globe given as latitude and longitude. However, due to a design mistake, it only recorded the directions of movement it had calculated—left and right—not the original data used to determine them.

This oversight meant that I could not trace the error back to the specific input—the exact location data—that caused it. Since the system appeared to work normally when I checked it, I could not recreate the problem to figure out what went wrong. In simpler terms, I was left without a way to understand the root cause of the incorrect direction being reported.

Losing the raw data removed any ability to re-evaluate, reapply or update that evaluation.[61] I could not test algorithm changes and potential fixes by replaying the raw data, and observing whether the output was then correct. Such scenarios in my professional life mirrored earlier lessons from my *Oma*, where she would stress that the conclusions of others require some critical scrutiny of the information from which they were derived.

Retroactive Policy Changes

How does the distinction between data and data analysis apply to identity? The famous author Mark Twain once likened data to

---

[61] This is true of algorithms that behave as something called a 'one way hash'. Such algorithms process an input to produce an output but cannot use that output to reconstruct the input, even if the processing steps are known. Imagine this as a calculator that only adds numbers. You may know that the calculator processes inputs by adding them together, and you may also know the output of a particular addition, but you will never be able to reverse engineer which specific input numbers were used to produce that output.

'garbage', highlighting the importance of collecting it only with a clear purpose. By contrast, officials in Nazi Germany collected data indiscriminately, with its purpose decided post-collection. (4 p. 87) This approach exposed people to unforeseen risks that apply equally to modern digital registries, where the future interpretation and consequences of data remain unclear.

The situation is perilous where a government decides the purpose for data, or who can access it, only after its collection. Australians can think back only a few years to our own census, and the depth and breadth of its questions. Information from religion to past military service was demanded under threat of penalty. (46) Governments come and go, but data remains, to be stolen, sold, or re-interpreted, indefinitely. These risks have historical precedents where laws were changed with profound implications for individuals and communities.

Comparing Historical and Modern Contexts

Consider the financial implications for Jewish religious communities in 1930's Germany when their public corporation status was revoked. An internal Nazi report from 1938 lamented the loss of tax revenue due to the dwindling number of these communities, and their deteriorating economic conditions. In response, the government enacted legislation on March 25, 1938, which:

> ... took away their rights as public corporations and made them associations with retroactive effect from January 1st, 1938. (23 p. 63)

This reclassification resulted in increased tax burdens, which in Berlin alone led to a budgetary rise of approximately 1.5 million Reichsmarks. The further back they applied the new law, the more tax they were owed.

Are retroactive policy changes relevant today? Combining digital ID with punitive measures for past actions can severely disrupt modern lives. The durable, immediately available data these systems collect record every minor action, turning even small transgressions into easily targeted future offenses. This both simplifies enforcement of liquid standards and amplifies the consequences of trivial issues.

Were you someone who diligently followed the often-erratic COVID era travel limitations? In Australia, for example, many people were restricted by police to a five-kilometre radius from their homes, with compliance driven by the fear of being branded a 'grandma killer', the pejorative label of the time.[62]

Now, imagine a scenario where this restriction is retroactively reduced to three kilometres. The motivations for this could range from increasing fine revenue, like we saw in Nazi Germany, to potentially encompass other, darker political agendas. These changes could criminalise once-legal travel, labelling people who ventured four kilometres from home as 'grandma killers' under the new standards.

If authorities have digital data to prove this, like phone location or facial recognition histories, they could easily penalise you for actions that were legal when you performed them. In contrast, had only the evaluation of your conduct been recorded at the time, that of simply 'being compliant', then any future change to the way this compliance is determined cannot brand you a criminal retrospectively. Pause to consider the potential risks associated with storing detailed personal digital data.

A 'Real World' Example

The concept of re-interpreting data for vastly different purposes is not far-fetched. In March 2023, it came to light that the Centres for

---

[62]   There were global social campaigns using this exact verbiage, spanning government departments such as the UK Department of Health and Social Care (99) and private creative agencies such as Open (98).

Disease Control and Prevention (CDC) had purchased location data for fifty-five million mobile users, for a home-by-home analysis of past curfew and lockdown compliance. Computer matching this data with other datasets provides detailed insights into personal behaviour, effectively placing citizens under warrantless surveillance, akin to parolees with ankle monitors. (47)

Simply misusing a stolen phone by exploiting its digital location, call history, and QR scans can falsely implicate others in crimes under this system. As reliance on technology's perceived accuracy increases, proving innocence becomes more challenging, underscoring the urgent need for strict data discipline.

Do we need to imagine such extreme hypotheticals? Retroactive policy changes still happen today. Take Australia's unclaimed money laws which allow funds from dormant bank accounts to be claimed by the government. In 2012, the inactivity threshold was cut from seven to three years, instantly reclassifying accounts idle for three to seven years as unclaimed, and at risk of seizure. (48) This affected various people, including those saving for home deposits and planning for retirement.

Limiting Identity Data Retention

Retroactive policy changes regarding personal identity can be dangerously destabilising for society. They can create legal complexities by potentially subjecting current activities to future, unforeseen laws. While they can address past wrongs, such as unpaid wages or legal appeals, they also risk unjustly altering the legal status of past actions.

This is unavoidable with digital identity data. Its properties make reliable deletion nearly impossible, rendering it unsuitable for centralised identity registration. Identity data is particularly sensitive due to its extensive and unbounded nature. This differs from trial evidence, which is inherently limited to case specifics and thus less prone to misuse.

It is hence crucial to submit identity data only selectively to digital systems from the outset. In Chapter Three, this exact approach protected our loan applicant, Jane. Because Jane owned the data being used to determine her loan eligibility, it was in her interest to only share the outcome of that determination with lenders, the analysis of the data, rather than the data itself.

A stark example of the risks associated with identity data retention comes from Nazi-occupied Poland, where census data, initially collected without malicious intent, was later exploited by occupiers to target Jews and others by identifying their religion and language. (4 p. xi) In contrast, if the Nazis were only able to access courtroom trial data, they would likely not have obtained information about defendants' religion, as that information typically falls outside the scope of most trials.

Digital Influence Over the Influential

Is anyone immune to the impact of digital systems? The enduring nature of digital data presents both opportunities and challenges, even extending its impact into the lives of the powerful. A striking example is found in the online *Bundesarchives*, which house a vast collection of digitised and publicly accessible German historical documents.

Among these, a particularly revealing case is a telegram from Dr Diego von Bergen, the Nazi Ambassador to the Church, concerning the Pope's upcoming speech at the 1933 College of Cardinals' Christmas reception (Figure 15). Labelled *"Ganz Geheim!"* [top secret], this document reveals the depths of government influence over societal narratives at the time, showing that even the Pope's words required Bergen's prior approval to align with Nazi principles. (49 p. 237) In Bergen's words:

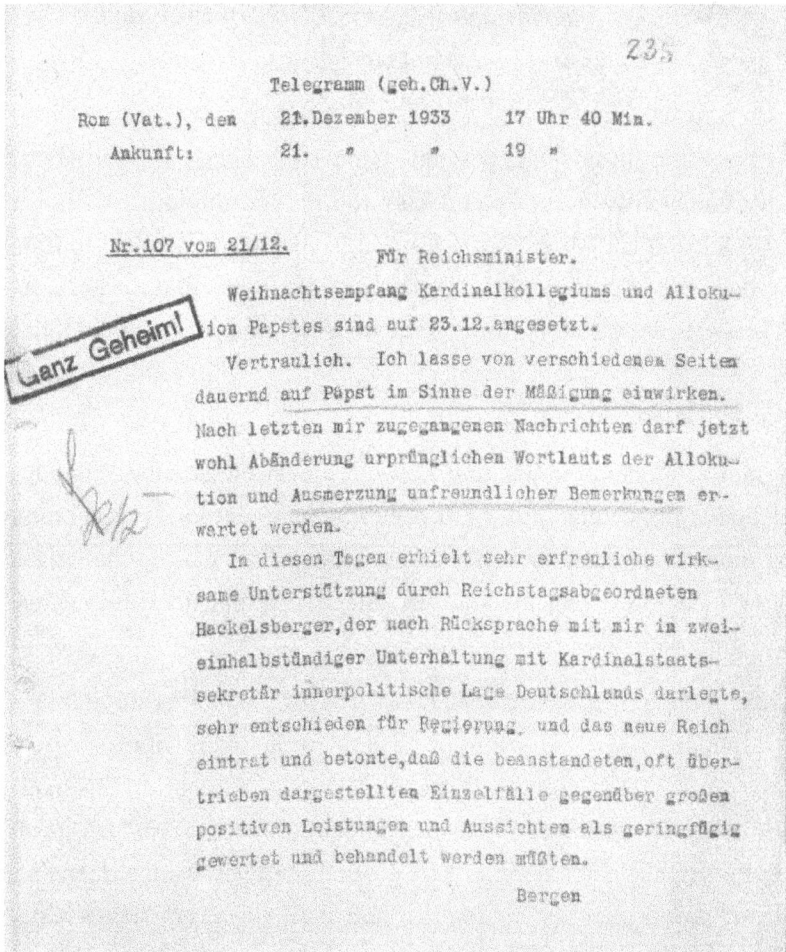

Figure 15. *Telegram from Dr Diego von Bergen, the Nazi Ambassador to the Church, concerning the Pope's upcoming speech at the 1933 College of Cardinals' Christmas reception. (49 p. 237)*

I am constantly influencing the Pope from various sides to encourage moderation. According to the latest news I have received, we can now expect changes to the original wording of the allocation and the elimination of unfriendly remarks...

... He spoke very firmly in favour of the government, and the new Reich, and stressed, that the individual cases

criticised, often exaggerated, must be considered, and
treated as insignificant compared to the great
positive achievements and prospects. (49 p. 237)

At the time, this document's top-secret classification shielded the government from scrutiny and kept the public uninformed. Had it been left in printed, analogue form, it would have been lost to time by now. Yet digitised, this document, once dangerous to possess, is now freely available online, revealing Bergen's once confidential words to the world and demonstrating the irreversible transparency of digitisation.

Bergen's telegram shows that even the most confidential communications from influential figures can become public, transcending the normal boundaries of space and time. This reality challenges the effectiveness of legislation in safeguarding against the misuse of digital data.

From communications to costumes, digital data haunts the influential in many ways. For example, in 2005, a photo of Britain's Prince Harry in a Nazi uniform made headlines across the world. This modern example illustrates the lasting impact digital records can have. Digitised, this image permanently documents Harry's misjudgement. He acknowledges this incident as "one of the biggest mistakes of his life" underscoring how digital records can perpetuate a momentary lapse beyond a private event, where behaviour may be viewed more leniently. (50)

We must carefully manage our digital footprints to protect our legacies. In today's digital age, controlling the storage and sharing of our personal data is essential to safeguard our narratives and identities. The alluring, yet enduring nature of online data has frequently caught even high-profile individuals off-guard. Let loose, digital ID risks perpetual immortalisation of our data, making Ambassador Bergen's and Prince Harry's embarrassment a daily reality for us all.

Data Sources: Pulling versus Pushing

Does digital technology change how data is collected as well as how it is stored? Data collection under the Nazi regime relied on a wide array of sources to compile identity records. This included census forms, employment details, church baptism records, and even academic research, such as PhD dissertations, to trace Jewish intellectual heritage. The aim was to seek out as much identity data as possible, a task undertaken by researchers committed to the Nazi cause.

The geographical scope of this data collection expanded as the Nazis occupied new territories. Yet under this system, data collection was limited to what could be actively found in, or 'pulled' from, the environment, restricting the scope and depth of information available. Now, the paradigm has shifted as people frequently upload, or 'push', their unsolicited data to online platforms via consumer devices and applications. This significantly broadens the data accessible to governments.[63]

Data-pushing is particularly concerning when it comes to children, exemplified by schools mandating the use of tablets that submit student activities to 'The Cloud'. This raises critical questions about the ownership of children's data, including their academic progress and personal thoughts. With government-controlled digital systems between us and our children, documenting the next generation's development, does the adage 'possession is nine-tenths of the law' hold true?

Implications For Digital ID

The pervasive nature of digital data, coupled with the assumption of uninterrupted connectivity, poses challenges when applied to iden-

---

[63] Data potentially accessible to governments is down to the granularity of live heartrate, cadence, and speed information from fitness monitoring devices. There is little more information that is actually measurable, let alone traceable.

tity management. These include technical failures, cyber-attacks, natural disasters, or deliberate disruptions, all potentially denying access to essential services requiring identification.[64] In addition, access to digital services is often limited in remote communities due to the high costs associated with their implementation.

Centralised digital ID also poses the risk of service revocation by operators for violations of seemingly unreasonable terms. Digital identification relies on software controlled by governments or corporations and requires their continuous permission for access. In contrast, physical IDs, like driver's licenses, offer resilience by operating independently of remote systems, dependably available in back-pockets across different emergency scenarios.

Recent incidents of 'de-banking', the unjustified termination of customer accounts leaving them unable to participate in the economy, underscore the importance of retaining non-digital payment options to sustain economic participation during service disruptions. Individuals with access to cash can continue transacting even without online capabilities, providing a dependable fallback. Digital ID systems suffer the same vulnerability.

The ease of duplicating and editing digital data, especially biometrics including fingerprint and retina scans to unlock devices, also adds significant vulnerability, inviting unauthorised duplication and alteration. While digital manipulation can easily implicate people in crimes, analogue identity documents from the past, like those used by the Nazis, were challenging to forge. Creating fake documents required labour-intensive efforts in candlelit basements and produced a relatively limited number of imperfect copies.

The enduring digital footprint of our identity casts long shadows over our personal legacy. As time progresses, physical documents

---

[64] Digital access has been a real issue during Australian bushfires. People resorted to stealing petrol from services stations because communications infrastructure had been destroyed in the fire rendering EFTPOS cards useless. Those with local access to cash could still transact as they were not reliant on accessing resources on remote digital systems.

begin to deteriorate, mirroring our own aging process. Yet digitisation halts this natural progression, indefinitely preserving our private moments. It is important to reflect on how digitised youthful mistakes could be perceived by future generations, friends, employers, or even by our future selves.

The idea that someone should permanently suffer for past actions is exemplified in the film, *Inglorious Bastards,* where American Lieutenant Aldo Raine carves Swastikas into the foreheads of captured German soldiers, forever scarring them with their wartime identities.[65] (51) Digital ID creates a similar permanent record of our actions, thoughts, and mannerisms, potentially exposing us to ongoing judgment. While seeking justice for war crimes should be timeless, these systems indiscriminately deny redemption for less extreme actions, and not one of us has failed to benefit from a second chance at some point.

A Better Future is Already Here

Building Online Reputation Today

Must we only dream about a different future? No, the use of technology to safely manage identity, as described by the trust agent model in Chapter Three, is more attainable than it may seem. Take Solana ID, a German start-up founded in January 2023 by industry experts.[66] Solana ID offers a product that:

> *… enables people to build a measurable online reputation and leverage good behaviour on the world wide web. (52)*

This platform meets the demand for verifying online behaviours, laying the groundwork for correctly managed institutional trust. Users can link their online activity, which Solana ID evaluates to generate a reputation score. This can accommodate social media engagement, financial transactions, and professional achievements, and is

---

[65]   American Lieutenant Aldo Raine played by Brad Pitt.

[66]   Solana ID, https://www.solana.id/.

enhanced by considering real-world credentials such as diplomas, making the assessment more accurate and relevant.

Solana ID serves as a trust agent by assessing the reputation scores of potential customers or partners with their consent. By focusing on reliable reputation scores rather than government evaluations of identity traits, Solana ID is supporting a shift towards a reputation-based society, contributing to a broader movement in this direction. (53)

Another existing offering comes from cheqd.[67] Their privacy-preserving payment network empowers individuals and organisations to control and monetise their own data. Their solution can be applied across various use cases in AI, decentralised reputation, supply chain, finance, e-commerce, education, gaming, travel, healthcare, and manufacturing. Powered by the concept of self-sovereign identity, explored in the next section, cheqd provides:

> ... *first-of-its-kind payment rails for decentralised identities that can be proven with Verifiable Credentials. (54)*

Likewise, Polygon ID offers:

> ... *privacy focused tools to put users in control of their identity across every digital surface. (55)*

Hypersign also provides privacy focused technology to create:

> ... *a decentralised identity layer for the internet, giving users control of their personal data and identity, whilst digitally enabling trust for businesses. (56)*

Self-Sovereign Identity (SSI): A Framework for Independence

Unlike government options, these private projects are built on the concept of self-sovereign identity (SSI).[68] SSI is a digital ID frame-

---

[67] cheqd, https://cheqd.io/.

[68] There's a great introduction to SSI at https://cheqd.io/ssi/.

work that allows people to manage and share their data independently, without centralised control. It relies on a 'trust triangle' —comprising the individual (or identity holder), the credential issuer, and the verifier—to securely confirm credentials while preserving privacy. But can this work in reality?

Imagine yourself at an amusement park, about to board the roller coaster. You are required to verify your height and age to do so but prefer not to reveal your full ID with sensitive details like your address and birthday. SSI functions like a special ticket, confirming only that your height and age exceed the requirements, not what they actually are. You hold the power to present only the information essential to the ride operator, safeguarding your other personal information.

Zero Knowledge Proofs (ZKP): Enhancing Privacy

Engineers use advanced mathematics called Zero Knowledge Proofs (ZKP) to achieve this. ZKP is a clever method which lets people prove they have certain information or qualifications without showing the details. For example, you can confirm your age or that you have a degree without giving away your birthdate or showing your diploma.

In the amusement park scenario, ZKP enables the ride operator (the verifier) to challenge you (the identity holder) to grab a ticket from a high shelf (the credential issuer) rather than measuring your height directly. Successfully retrieving the ticket, which is placed at the ride's minimum height requirement, serves as sufficient proof that you are tall enough. This act of demonstrating suitability is analogous to how King Arthur proved his entitlement, without relying on external identification.

Demonstrating that you meet specific criteria without disclosing the precise details that invite identity related risks is the essence of

ZKP. Here, the ticket acts as the proof of your eligibility, assuring the operator of your qualification without an actual measurement, thereby protecting your privacy.

The Practicality of Privacy-Centric Tools

Grasping the technical details behind privacy-centric products is not required, in the same way that people don't need to understand how Smart TVs work to stream movies. The key is that these products work smoothly, offering benefits without the need to understand the underlying complexities.

The potential applications of this approach are vast. Imagine the convenience of anonymous home rental recommendations, meeting all necessary checks—references, rental history, financial suitability, and tenancy behaviour—without revealing sensitive information like employment and salary details. This technology can revolutionise many activities, from housing to voting and online grocery shopping, while effectively removing the risks of government involvement.

Chapter Takeaways

This chapter has illuminated the potential perils of government involvement in digital ID technology. Here, we explore actionable advice to help protect and value our identities in the digital age, drawing on the insights shared throughout the chapter.

In Private

Practice Digital Minimalism

Selective engagement with technology allows for a more controlled digital footprint, mitigating the risks of data misuse. Be mindful of the digital tools and services you use and opt for non-digital alternatives when possible. Digital minimalism involves more than cutting down on gadgets and apps; it is about thoughtfully curating your online presence to prevent data misuse. Here are some practical tips:

- Be selective about the social media platforms you engage with, the content you share, and the personal details you disclose in your profiles.

- Avoid using voice-operated smart-home devices, which can collect data on your personal habits.

- Store photos on external drives instead of the Cloud.

- Disable location services on your phone and opt for physical maps or street directories for navigation.

- Use a physical diary or calendar instead of a digital one. It is not only secure but also provides a tangible record of your year that is enjoyable to revisit.

- Avoid digital driver's licenses and other seldom used digital credentials or memberships, like library cards, when physical ones are available and practical. Recent evidence suggests that these government systems are plagued with issues.[69]

- Review and manage the permissions you grant to apps and services regularly.

- Use fewer services and avoid reusing usernames and passwords to ensure that credentials in any one system, if exposed, cannot be used elsewhere. This industry best practice, incidentally, directly contradicts current political advocacy.[70]

---

[69] In November 2023, Queensland, Australia, launched a digital driver's license system that encountered errors, despite an AUD $53 million investment and half a decade of development. (102) This incident echoed a similar issue in Victoria, Australia, in June 2023, where the digital driver's license system mistakenly sent 57,000 emails with incorrect surnames to drivers. (101)

[70] A press release from Australian Senator Katy Gallagher on March 27, 2024, claimed an advantage of centralised government digital ID is to "reduce the need to remember many different usernames and passwords for different services by providing a reusable digital ID that can be used instead." (96)

- Turn your phone off at night to reduce circadian rhythm profiling, and incidentally, get a better sleep.

Regular digital detoxes and periods of disconnection can further enhance your privacy by allowing you to assess and streamline your technology use. Inspired by the risks of technology-driven surveillance, adopting a minimalist digital lifestyle fosters a healthier, more purposeful interaction with technology, ensuring it benefits rather than detracts from your quality of life.

Embrace Privacy-Enhancing Products

Using privacy-enhancing technologies is essential in today's digital landscape. Virtual Private Networks (VPNs), such as Virtual-Shield® or NordVPN®, secure your internet activity by concealing your device's details. Choose encrypted messaging apps like Signal over SMS to protect your conversations legally. To further safeguard your privacy, intentionally misspell words or use numbers and symbols in your messages, making it harder for surveillance software to interpret your communications.

Consider switching to paid online services that prioritise privacy more than free ones. Companies need to finance their operations, and without product revenue, monetising user data is often their strategy. Privacy-focused email providers like StartMail or Proton-Mail offer enhanced security and are less accessible to unauthorised parties. A useful guideline to remember is that if you're not paying for a product, you likely are the product.

After migrating your existing products to more secure alternatives, focus on future developments. Learning about platforms like Solana ID, cheqd, Polygon ID, and Hypersign is a great place to start. Engaging with the entrepreneurs behind these products signals market acceptance, encouraging them to continue their important work. Offering authentic feedback on user experience is crucial for making these products accessible to the general public.

In Public
Practical Tips for Legally Avoiding Prying Eyes

Surveillance cameras are pervasive in public spaces, monitoring everything from vehicle license plates to pedestrian movements. To enhance your privacy against these prying eyes, consider the following strategies:

- Obscure your face at ATMs or other automated machines.

- Avoid directly gazing into overhead cameras.

- Wear sunglasses and hats to conceal your facial features.

- Change your walking speed to confuse gait recognition systems.

To maintain privacy while using digital navigation, a clever tactic is to input an address close to your destination, perhaps a few numbers away. This small adjustment can lead to a nearby residence or business that has no connection to you, concealing your actual location and affiliations, while benefiting from the convenience of online maps.

Also consider avoiding toll roads with cameras and electronic tags, as these systems monitor your movements and link to your financial transactions. Opting for alternative routes can also be cost-effective. Whenever possible, use paper tickets for public transport instead of digital alternatives to further reduce your ability to be tracked. By implementing these strategies, you choose to use technology as a liberator rather than an oppressor.

A Choice

This chapter has examined the impact of technology on human identity and has led us to a critical juncture. We now face a digital ID arms race between the empowering prospects of free-market solutions, and the control and surveillance of centralised government alternatives.

Yet technology is not the villain of our story; it simply mirrors our collective choices and values. Now is the time for each of us to assert these values. Developers: innovate with ethics at the forefront of your designs, ensuring that privacy and autonomy are not afterthoughts. Policymakers: legislate with a vision that prioritises individual rights and fosters an environment where decentralised technologies can thrive. Citizens: expect more from your officials in the identity space by demanding they do less.

The future of technology in identity management hinges on its ethical and restricted use, not its rejection. Yet a free future comes with a crucial caveat: the exclusion of government control over digital ID. When governments assume this role, the potential for abuse is not purely theoretical, but historical fact.

Today's politicians continue to erode our privacy through new laws. Take the recent updates to eIDAS (electronic identification and trust services), the EU's online identification regulation, for example. Article 45 grants the legal authority, but not the technical ability, to monitor SSI users (57). In the US, the impending Real ID Act, effective May 7, 2025, represents a move from state-issued IDs to a federally controlled system, restricting certain uses of driver's licenses for air travel and federal building access. (58)

These changes require engineers to build the systems that support them, coding potential back-doors that compromise anonymity, and allow bureaucrats into the gates. This brings us to the most dangerous component of digital ID: government.

CHAPTER 5:
Government

We would have been at a loss if they had disappeared before being registered and concentrated.

—Obersturmbannführer (Colonel) Adolf Eichmann, Senior Nazi SS officer,
co-architect of the Holocaust, answering the question,
"How could the Jews have resisted?" hours before his execution,
1 June 1962. (4 p. 92)

Privacy is a vulnerable value in the face of demands for administrative efficiency and attractive estimates of revenue gains.

—Joint Select Committee Report on the Australia Card proposal,
May 1986. (12 p. 113)

Potentially, a government is the most dangerous threat to man's rights: it holds a legal monopoly on the use of physical force against legally disarmed victims.

—Ayn Rand

A hostile informal system lurks around every step, every word, and every official act. More than dubious elements try to make themselves important to the authorities as saviours of the allegedly threatened state.

—Memorandum to the Nazi government from His Excellence, Cardinal Pacelli,
written to Ambassador Dr Diego von Bergen. Key discussion points for the
upcoming meeting on recent church-government tensions, 13 Jan 1934.

A picture can tell a thousand words. This picture of my German great-great-grandparents was taken on holiday and sent as a post-card to their daughter, my *Oma's* mother, still a schoolgirl at the time (Figure 16). This was in August 1904, many years before the Nazi Government's rise. When this photograph was taken, a form of identity registration was already in place in Germany and registry offices had been storing basic identity data, including religion, for a staggering twenty-nine years. (4 p. 73)

Figure 16. My German great-great-grandparents on holiday (1904). This picture was sent as a postcard to their daughter, Charlotte, my great-grand-mother.

Yet this picture's innocent recipient was oblivious to this information. She had no idea that the identity data held by the pre-Weimar government of the time would be weaponised by an entirely different one, decades later. Nor could she have known that future politicians' policies would force her to protect her daughter, my *Oma*, from approaching gunfire in a Berlin basement forty-one years later. Today, that 120-year-old family photograph is a reminder of the timeless risk of data in changing government hands.

Our discussions to date have laid a foundation for grasping the fusion of identity, trust, and technology. In this chapter, we unveil the core danger of digital ID: centralisation under government control. Using Nazi Germany again as a backdrop, we examine the essence of centralised power and its significance in identity management.

This critique does not only highlight the failings of centralised identity management, but also of government, when it adopts that central role. Unlike private businesses, which can misuse identity data by denying services, governments possess the unique threat of military and police enforcement. When democratic oversight fails, centralised identity registries lack safeguards for public protection. In the hands of a rogue government, atrocities may be legalised by the stroke of a dictator's pen and implemented by their power-hungry henchmen.

When detailed personal data is readily available to officialdom, those labelled as undesirable are left with no redress, legal or otherwise. Letting a government with a monopoly on force, know everything about you, is a potential recipe for disaster in an era characterised by popular psychologist, author, and commentator, Dr Jordan Peterson, as one where:

> *... we're at odds with one another about identity.* [71] *(59)*

---

[71] Closing remarks at the inaugural Alliance for Responsible Citizenship (ARC) Conference in October 2023 held in London, UK.

The Nature of Government

The political class holds the unique authority to determine legality, and exercise powers denied to the wider population. Governments, often seen as central figures of societal benefit, have increasingly become the sole decision-makers in society. Yet, as highlighted by the Byzantine General's Problem in Chapter Three, centralising control tends to shift, rather than solve issues, potentially creating larger problems. This centralisation also burdens decision-makers with high expectations, commensurate with their power. Regardless, they enjoy disproportionate influence.

Governments lack the credentials to qualify for this auspicious role. They are not, for example, comprised of superhumans, immune to temptation and corruption. Nor are they necessarily staffed by experts in philosophy, history, technology, or economics. All of these are fields conducive to healthy societies.

Regardless, this imbalance is pervasive. For example, law enforcement officers are legally permitted to carry firearms, exceed speed limits, and detain people—actions that attract punishments for civilians. Similarly, conscription can force people into harm's way against their will, a demand not placed on individuals by their peers. Additionally, the government's authority to create currency and impose taxes would be defined as fraud and theft if performed by commoners.

How have organisations that perpetuate such inequality become so deeply embedded in society, often with widespread approval? One explanation is the so-called 'cloak of democracy', which suggests that voting gives citizens the illusion that "we are the government." This belief implies that government actions are self-imposed, making it difficult to protest against our own choices. However, the idea that a majority-elected government is truly representative is questionable.

Economic historian and political theorist, Murray Rothbard, highlights the fallacy of such a claim:

Under this reasoning, any Jews murdered by the Nazi government were not murdered; instead, they must have committed suicide, since they were the government (which was democratically chosen) and, therefore, anything the government did to them, was voluntary on their part. (60 p. 10)

What is the Government?

If we aren't the government, then who truly holds power? Before we entrust our most valuable asset—our identity—to such an omnipotent entity, we must thoroughly scrutinise its nature. Rothbard offers one definition for this centralised bureaucracy, often presented as the panacea for society's digital ID management.

That organisation in society which attempts to maintain a monopoly of the use of force and violence in a given territorial area. (60 p. 11)

German Jewish political economist and sociologist Franz Oppenheimer offers yet another definition:

The systematisation of the predatory process over a given territory. (60 p. 15)

I also, humbly, offer my own:

The methodisation of moral exceptions by a minority, seeking the acquisition and preservation of power over humanity.

These confronting definitions demand justification. Despite governments often masking these characteristics under a flimsy guise of civility, our previous discussions on labels encourage us to look beyond the surface. By applying healthy scepticism towards government actions, we can truly understand them by their deeds.

How Do Governments Preserve Their Power?

Leaders across all forms of government, from dictators to democratically-elected officials, depend on community backing to sustain their authority. No single person can control a nation without support. Gaining public approval is hence essential for any politician looking to maintain power and enact laws affecting entire populations.

Governments lack the resources to directly enforce behaviour-regulating policies. Instead, they rely on the threat of punishment as a deterrent to encourage individuals and organisations to self-enforce their doctrine. The success of this approach hinges on maintaining authority, essential for enforcing social ideologies and, co-incidentally, generating profit, even post-tenure. Those who benefit from this system are therefore naturally inclined to focus on preserving their advantageous positions.

Governments often employ tactics such as selective resource allocation, information control, and legislative measures to maintain power, reward supporters, and suppress opposition. These strategies can include manipulating elections, influencing the judiciary, and declaring emergencies to shape public opinion and secure incumbency.[72, 73]

If we view government, even partly, as a power-seeking institution, shouldn't we re-evaluate its decisions in the context of this incentive? To determine whether government initiatives, like digital ID, truly benefit the public, or simply bolster bureaucratic power, we need to scrutinise the three key aspects of government self-preservation— economic, ideological, and emotional.

---

[72] The design of electoral systems, including gerrymandering and campaign finance laws, can be manipulated to favour incumbents.

[73] These techniques are often implemented to assume crisis powers.

Economic Means

Control is often maintained by economically incentivising certain groups, selectively sharing the spoils of power to garner political support. This *quid pro quo* often comes at the public's expense, as governments effectively purchase loyalty, and capabilities, from the influential.

This strategy was seen in the Nazi regime's symbiotic ties with companies like *Dehomag*, where government funds were traded for technological support. A parallel existed with their domestic enforcement agencies, power-hungry opportunists sustained by taxes and currency creation, in exchange for suppressing dissent in the streets—a practice perhaps reminiscent of some modern police forces. We will revisit these agencies later in this chapter.

Modern governments still spread their ideologies using society's products and services. But what can governments offer in return? Providing reciprocal value via taxation and currency creation has limits. Information, on the other hand, is limitless, inherently valuable, and can be monetised externally.

Chapter Two revealed that information on population identity is particularly valuable as a key to unlock human resources. This casts new light on governmental ambitions to control digital ID. These systems not only offer the government an economic asset that can be leveraged but are simultaneously promoted to the public as a societal good. A government monopoly on digital ID would become such a lifeblood that it creates a conflict of interest, disqualifying it for the role.

Ideological Means

Can we really trust the experts? Governments maintain power by projecting an image of benevolence and integrity, largely by leveraging media to promote narratives from compliant intellectuals.

These influencers constitute an informal aristocracy and craft narratives that steer public opinion in favour of government doctrines.

This approach, known as the ideological means of state preservation, is seen in the rise of fact-checking services. The modern mantra 'trust the experts' encapsulates this strategy.[74] Many find critical thinking challenging, making clear the allure of deferring societal judgements to authority. Relying on the views of so-called experts, frequently financed by governments, offers a convenient intellectual shortcut.

How was the ideological means of preservation used in Nazi Germany? Joseph Göbbels, the Reich Minister of Propaganda, and once an unsuccessful misfit, exchanged his skill in crafting pro-government narratives, for prestige and power within the regime.[75] His articulate mastery of labels and slogans, as detailed in Chapter Three, arguably pioneered the methods used to suppress scepticism towards government policy. Göbbels, whose words haunted my *Oma* in that Berlin basement, famously proclaimed:

If you repeat a lie often enough, it becomes the truth. (15 p. 63)

One such lie was that citizens must forfeit their identities to the government for national well-being, even in trivial matters. To this end, the Nazis skilfully used 'social engineering', a technique identified by philosopher Sir Karl Popper, for implementing major changes under the pretence of an:

... alleged greater good. (12 p. 176)

---

[74] And, incidentally, this strategy highlights the abuse of the word 'trust' discussed in Chapter Three.

[75] Göbbels held this position from 1933 to 1945 having been previously unemployed. Physically handicapped and unable to make a living through his academic pursuits (which included writing and teaching), he was a professional failure before Hitler appointed him as reward for the use of propaganda and vitriol in recruiting new party members. Such are the impacts when cronyism displaces meritocracy in government.

In the 1930s, through persuasive slogans, orchestrated events, and media control, including cinema, they spread their ideology, persuading the public that Germany's reconstruction was more important than, and even at odds with, individual rights and privacy. This narrative, reinforced across all media and with dissent criminalised, created an illusion of unanimous ideological support with modern similarities.[76]

The Nazis pushed the idea that the more they knew about their citizens, the better they could serve and protect them. Initially sold as a means to catalogue skills for national development, with an emphasis on female fertility for re-population, they pushed social, technical, legal, logistical, and statistical boundaries, to collect personal data extensively. However, as introduced in Chapter One, Friedrich Burgdörfer ultimately exploited this data to plan the Polish invasion and enhance the *Volkskartei*.

As effective as they are, economic and ideological measures cannot suppress the truth indefinitely, even in the most powerful regimes. A single act of dissent can ignite widespread critical thinking. Rothbard even identified independent intellectual criticism as "the greatest danger to the State." (60 p. 25) Thus, the most effective way for a government to maintain power is by manipulating public emotion.

Emotional Means

Humans, as emotional beings, are easily swayed by emotional manipulation. Pro-government intellectuals can exploit emotions such as guilt, to foster self-censorship among critics. By portraying opposition as selfish, and against the common good, they effectively discredit divergent thinking and encourage conformity.

---

[76] This is similar to the relentless censorship of doctors and professionals providing clinical evidence supporting narratives contrary to the officially sanctioned government narrative of COVID. Interestingly, many of these opposing opinions are now being proven correct.

Emotionally loaded wording is prolific. For example, tax breaks for the wealthy are frequently condemned as pandering to the rich. Likewise, COVID era vaccine sceptics were branded as selfish public health risks. In identity management too, scepticism is stigmatised, insinuating that critics have something nefarious to conceal.

Emotional manipulation was a key tactic in Hitler's rise to power. In the 1930s, the government sought public compliance with identity management through emotional appeals. The era following the First World War, marked by economic devastation from the 1919 Treaty of Versailles—seen by Germans as a betrayal—provided this opportunity. The treaty stripped territory and imposed fines, inflicting national humiliation and a sense of impotence upon Germans. (3 p. x)

Article 231, the 'War Guilt' clause of the treaty, saddled Germany with heavy reparations, triggering hyperinflation and societal unrest. Amidst this chaos, Hitler capitalised on the nation's despair, presenting himself as a saviour, and manipulating events to further his ambitions.[77]

With strategic cunning, Hitler crafted a mass extermination plan under the cloak of democracy, systematically taking control of the government, media, and corporations to legitimise his heinous acts. Using his charisma and grand ceremonies, he ignited patriotism and self-interest, promising to rejuvenate the aggrieved nation. His ascent was fuelled by a wave of renewed optimism, as he mesmerised the public with compelling promises through relentless social engineering.

Yet, the promise of prosperity was merely a lure. Hitler also wielded the potent motivator of fear. The government stoked panic over vague threats, labelling criminals and 'anti-socials' as obstacles to the promised utopia. Claiming the need to confront these fabricated

---

[77] Even among the Allies, concerns arose regarding the severity of the Treaty because of its potential to encourage radical leadership in Germany.

dangers, justified invasive anti-privacy laws, ostensibly to identify and neutralise threats to German prosperity.

The public was convinced that on-demand identification was essential for safety, given the alleged omnipresence of enemies. Critics were marginalised and their dissent criminalised. Any opposition to identity collection became a life-threatening act. Nazi Germany epitomised tyranny, deeply embedded within its legal framework. In particular, the *Reich Registration Order* of January 6, 1938, was presented under the guise of "the protection of the people against criminals and the Security Police's fight against those criminals." This is a narrative that, arguably, is still alive today. (4 p. 38)

Hitler's ascent capitalised on the economic, ideological, and emotional pillars of government power, notably using the grandeur and theatre of the Nuremberg rallies, to foster unity among Germans in their despair. Yet, this unity demanded a steep cost: the relinquishment of personal freedoms—a sacrifice many deemed necessary for the greater good. State-driven social engineering clouded their judgment, prompting them to exchange their liberties for alluring, yet ultimately ephemeral promises.

Government Impacts on Decision Making and Trust

Trust should be grounded in transparency and accountability, rather than presumed inherent integrity. While some citizens may trust their government's ability to handle identity data, such trust is not universal, and can be misguided. Like the Byzantine Command Centre we discussed in Chapter Three, governments are centralised bodies made up of fallible individuals; they are susceptible to corruption. History shows that misplaced trust in such institutions can lead to exploitation, with governments sometimes acting against their citizens' interests.

How important is a high standard of transparency in governmental communications? Our ability to make informed decisions in various aspects of life depends on access to accurate and transparent information. This capacity is undermined when information is distorted by those with vested interests in certain outcomes, potentially leading to erroneous conclusions.

This concept is widely recognised. For example, financial advisors may prioritise commissions over client needs. Likewise, governments that promote centralised infrastructures controlling information flow raise concerns about impartiality and fairness. Thus, policies like digital ID development should originate from grassroots initiatives rather than top-down government mandates. Incentives aside, the efficiency and challenges of government-managed public services, especially in identity management, must also be carefully considered.

Efficient Public Services
Are Governments Efficient?
A critical examination of the government's service delivery history is needed to evaluate its ability to implement digital ID. One viewpoint supports the advantages of centralised government identity management, suggesting it streamlines administrative processes and simplifies access to government benefits. However, this assumes government efficiency is always beneficial and that digital ID actually provides this efficiency.

But are government efficient? While government services can offer value, they often lack effectiveness. For example, following recent violent storms in my area, local initiatives swiftly provided generators, cleared roads, and assisted the elderly, significantly outperforming slower government disaster responses. This underlying inefficiency arises in part when solutions become distanced from problems and is made worse by government's focus on re-distributing resources, rather than directly producing them.

Anecdotally, complaints about government inefficiencies are widespread, sometimes even becoming a cultural pastime. This is commonly seen with local council issues like park maintenance or street lighting, which frequently ignite social media criticism. As government size grows, efficiency can decline, with the effectiveness of actions diminishing due to their increasingly indirect impact and isolation from immediate, local feedback.[78]

Over time, and despite increased budgets, government services often show little corresponding improvement in timeliness or quality. This suggests a point of diminishing returns, where additional spending does not translate into greater efficiency, and may even impede it. With government growth comes more bureaucracy and financial burden. The potential benefits of centralised services, like digital ID, must consider the mammoth costs to establish and maintain them, lest we fall into the trap of a false economy. We will revisit this topic in the next chapter's case study.

Are Governments Efficient at Appropriate Things?

The idea that government efficiency is always beneficial deserves equally thorough scrutiny. Although the label 'efficiency' typically suggests benefits, such as reduced costs or saved time, it can be harmful in certain contexts. In these cases, such efficiency can benefit the government more than the community it serves. Certainly, the Jews of Europe did not benefit from the efficiency of those hunting them.

A study from Nazi Germany challenges our views on government efficiency, particularly in the context of pre-war Jewish emigration,

---

[78] This is an issue possibly addressed by micro-issue voting, as already discussed.

where methods in two parts of the Reich were compared. Contrasting the so-called *Ostmark* and the 'Old Reich', emigration was severely impeded in the latter due to bureaucratic inefficiencies.[79], [80] However, in the *Ostmark*, the establishment of a Central Office for Jewish Emigration centralised and streamlined the process, allowing for quicker emigration with all required documents.[81] (23 p. 77) This example illustrates how efficiency can be manipulated for harmful purposes. The report states:

These abuses that occurred in the old Reich were controlled in the Ostmark by concentrating the entire organisational system on emigration and the establishment of a "central Office for Jewish Emigration" (August 26, 1938) under the direction of the Inspector of the Security Police.

This "central office" brings together all the authorities involved in the emigration of Jews, so that it is generally possible to provide Jews willing to emigrate with all the documents necessary for emigration within a period of 8 to 14 days. (23 p. 77)

In addition, the central office decides together with the Vienna Foreign Exchange office and the "Israeli religious community" about the distribution of the foreign currency assets made available by the foreign Jewish organisations, which made it possible for the emigration of poor Jews at the expense of wealthy Jews on a large scale, and in a planned manner. (23 p. 79)

---

[79] The *Ostmark* generally referred to Austria.

[80] The 'Old Reich' referred to the original parts of the protectorate.

[81] Emigration in the old part of the Reich, with the participation of the majority of Jewish political and religious organisation, was in the hands of the Reich Representation and the Aid Association of Jews in Germany that was integrated into it. Their responsibilities included worrying about procuring the foreign currency necessary for emigration. However, apart from group transport, the emigrant was responsible for procuring the emigration papers, including certificates of good conduct, official declarations of no objection, and passports.

The report tabulated the *Ostmark's* efficient emigration process, leading to proposals for its adoption in the Old Reich. Yet, this efficiency facilitated a harmful agenda, suggesting that government inefficiency might actually shield against abuses of power. In oppressive regimes, inefficiency could paradoxically protect against exploitation. Given the complexities and potential for misuse of power within centralised governments, it becomes imperative to examine how such entities prioritise and categorise policy issues they wish to streamline, distinguishing between those that are non-negotiable and those that allow for flexibility and compromise.

Absolute and Relative Policies

The Expectations of Categorisation

Understanding the policies that attract legislative interest brings clarity to decision-making integrity. These policies generally fall into two categories: absolute policies, which are non-negotiable, and relative policies, which permit compromise and balance between differing, justifiable perspectives.

This categorisation applies to many aspects of life. For example, ethics and philosophy explore moral absolutism—actions considered inherently right or wrong—and moral relativism where context influences morality. Similarly, risk management and safety engineering distinguish between absolute specifications, such as the weight-carrying capacity of a bridge, and those that allow optimisation within acceptable limits.

Absolutism and relativism are also visible in public policy. Managing public infrastructure, like roads and parks, involves relative decisions, balancing economic and social factors. Conversely, crimes like rape and murder are absolute issues. A functional society does not debate acceptable levels of these evils, but rather unanimously agrees on zero tolerance and flexibility.

Is determining the category of political issues always clear-cut? Mis-classifying them, whether deliberately or unintentionally, can result in undesirable and potentially devastating outcomes. Considering our knowledge of technology-driven identity management, the question arises: where does centralised government digital ID fit, and is its classification being respected?

Relativising an Absolute Issue

On initial inspection, consolidating identity management under government control may appear advantageous. However, debating this crucial issue risks applying relative thinking to an absolute scenario, potentially eroding vital social standards.

Consider the aviation industry, termed 'safety-critical' by engineers because failures can be fatal. Strict and complex safety protocols are in place to ensure reliability, with passenger trust built through absolute, albeit sometimes inconvenient, standards for equipment and processes. Part of these non-negotiable protocols demand that pilots are appropriately trained, sober, and well-rested, with no exceptions.

But what if these lines were blurred? For instance, a proposal allowing slightly inebriated pilots to fly, under the guise of inclusivity and stress reduction, could be disastrous. Proponents might argue that a small amount of alcohol helps certain pilots relax, potentially enhancing their performance and reducing flight delays by having standby pilots readily available at airport bars.

Pursuing this 'logic', should we consider eliminating expensive pilot licenses and training to combat financial discrimination? Clearly, this approach does not eliminate discrimination, but instead merely shifts it from those unable to afford to train, to passengers who bear the consequences of lowered standards. This absurd example underscores the risks of applying flexible thinking to inflexible issues.

Compromising on critical standards, such as a pilot's cognition, threatens the entire system, as these standards are essential for maintaining industry-wide safety.

This example illustrates that although the labels of convenience, equal opportunity, and reduced stress are initially superficially attractive, they should not lead to the compromise of critical safety standards. We must either uphold these non-negotiable principles or admit that compromising them risks public safety by normalising hazardous practices.

Digital ID must be considered safety-critical. Its implementation must meet the same strict standards as aviation, because failures can cause harm and death. Unfortunately, government regulation of this area is already challenged by conflicts of interest and impartiality concerns. Yet unlike aviation, where people who distrust the system can simply opt not to fly, abstinence is largely infeasible with government digital ID programs, especially when they become quasi-mandated. For individuals, it's akin to realising the pilot is drunk only after take-off.

Government Involvement in Identity
From Historical Roots to Digital ID Dilemmas

Governments have been encroaching on identity management for centuries. Some trace their involvement back to the Westphalian system, established in 1648, which affirmed government territorial sovereignty and lead to civil registration.[82] Others credit standardisation of the Napoleonic Code in 1815 which marked the start of censuses in German states.[83] (4 p. 14)

---

[82] This was born from the end of the Thirty Years' War.

[83] The Napoleonic Code was also known as the Civil Code of 1804, which standardised civil registration of citizens by their governments.

Some suggest there will be a place for government in identification for centuries still, with digital ID potentially augmenting services like healthcare. While it is true that digital ID could enable quick access to data like patient records, aiding emergency identification, it simultaneously opens a Pandora's box of concerns that bleed into other areas of government operation. These include the misuse of health data for discrimination or genetic profiling, reminiscent of historical eugenic practices.

The Holocaust is particularly instructive in this regard. Unlike other genocides, such as those in the Ottoman Empire, Cambodia, and Rwanda, the Holocaust was distinct in its meticulous execution through laws and bureaucracy. Although an extreme example, it lays bare the severe outcomes possible when government and identity management intertwine.

Erosion of Trust and Autonomy Under Surveillance

Many Germans living under Nazi rule tried to avoid the initially fragmented identity registration system, with some turning evasion into a sport. But as the government stoked fears of enemies, people increasingly looked to it for guidance on whom to trust. (4 pp. 146-147) This shift slowly dismantled the shared moral values foundational to German society.

Employment as a Means of Control

The workplace was a key area for identity registration, shown by the introduction of the *Arbeitsbuch* [workbook] by the Nazis on June 1, 1935. This document made worker credibility entirely dependent on government records. These could be altered or nullified without recourse, overshadowing individual reputation and merit.[84]

---

[84] Mirroring the *Arbeitsbuch*, in September 2023, the Australian government introduced the concept of a digital skills passport as an extension of its digital ID initiative, with Treasurer Jim Chalmers endorsing the project. (100)

Three weeks before the *Arbeitsbuch* was implemented, Martin Bormann, a senior Nazi official and one of Hitler's close confidants, issued a directive for bureaucratic efficiency: all labour-related inquiries were to be funnelled to a designated officer for internal party matters. (61 p. 5) This directive highlighted the Nazi regime's emphasis on administrative centralisation and efficiency, often at the cost of individual rights. Bormann said:

> *I have noticed that questions on the labour front are dealt*
> *with by a wide variety of staff departments. I order with*
> *immediate effect that all letters and inquiries concerning*
> *the labour front are to be forwarded to [party member]*
> *Friedrich's officer for internal party affairs. (61 p. 5)*

Yet the *Arbeitsbuch*, ostensibly a work history record, also served as a surveillance tool. It captured personal details such as nationality, interview notes, and signatures, as well as the number of children the holder was supporting. (62 p. 201) This dual function is shown in correspondence from a German police chief in 1942. His discussion of female workers from former Soviet territories illustrates the use of employment in surveillance and racial profiling:

> *I ask you to immediately contact the employment administra-*
> *tion recruiting offices located at these locations – if necessary, via*
> *the security police and SD[85] commands – to make the necessary*
> *arrangements and to carry out the screening when recruitment*
> *begins. (63 p. 14)*

> *It is only necessary to ensure that racial suitability is noted in the*
> *transport lists during the screening according to paragraph*
> *1.2 and, if possible, on the work cards.*

> *During the screening, according to paragraph 1.3,*
> *it would also be advisable to aim for the racial unsuitability*
> *to be recorded on the work card. (63 p. 15)*

---

[85]   The SD *(Sicherheitsdienst)* or security service will be covered later in this chapter.

When does mere monitoring cross the line to become intrusion? By 1944, employment served as a mechanism for the government to monitor pregnancies. Correspondence of August 17, which refers to fathers as mere 'producers', exemplifies the de-humanising intrusion possible when involving employers:

> *In his decree B II 402/44 of June 5, 1944, the Reich Minister of the Interior determined the treatment of pregnant foreign workers and children born in the Reich to foreign workers and clarified the procedure.*

> *Accordingly, the illegitimate pregnancies of the foreign workers are reported to the youth welfare office by the company via the employee, which then independently identifies the producers.*

> *I would like to point out that the question of the nationality of the producer is unimportant, whereas the nationality of the impregnated woman must not be overlooked. (64 p. 7)*

Re-classifying Citizenship and Political Rights

Just three months after the introduction of the *Arbeitsbuch*, the Nuremberg Laws were unveiled at a theatrical rally, marking a significant shift in human classification. These laws created a hierarchy of citizenship and abolished legal equality by assessing individuals based on their ancestry. Terms like 'full Jew' and 'mixed breeds' of the first and second degree resulted in conditional social rights.

Following the Nuremberg Laws, subsequent regulations like the *Health Pedigree Book* (1936) and the *Duty to Register* (1938) further diminished privacy and autonomy. These laws tipped the scales of knowledge in favour of the government, streamlining the process of identifying and arresting individuals deemed 'undesirable' as outlined by the Racial Political Office:

*All Germans whose membership of the German people is clearly
established must be included in a German people's list. You will
receive German citizenship. Only these Germans are entitled to Reich
citizenship. All other people have no claim to imperial citizenship,
and therefore no political rights. (20 p. 549)*

A pervasive fear gripped the nation, fuelled by uncertainty over
whether one's registered details could suddenly fit the criteria for
arrest. The unchecked power of the police enabled arbitrary targeting and assault of civilians, illustrating the 'might makes right'
doctrine upon which governments operate. This was a case of the
powerful preying on the vulnerable, which transformed once-civi-
lised streets into lawless jungles. My *Oma*, like many others, would
start each day asking an unanswerable question:

Am I undesirable?

Strategic Impacts of Identity Centralisation

The government increasingly misused identity data. Registration
had identified men of military age, leading to the reinstatement of
the draft on March 16, 1935. (3 p. 527) Mandatory Labour Service
was introduced in the same year, requiring recent school graduates
to spend six months in camps that ostensibly focused on forest
restoration, but actually imposed military training and doctrine.
(3 p. 21)

By November 1937, the *Status of Persons Law* mandated coordinated
registration efforts. (4 pp. 39-40) The next year, sharing identity data
across government departments became recognised as both innovative and essential. (4 p. 39) A Reich Security Main Office report
highlighted that centralising identity systems helped meet Jewish
emigration quotas in 1938. (23 p. 57) Additionally, during the occupation of Holland, there were demands for centralised access to registration data for all administrative bodies, including courts. (4 p. 67)

Concentration and death camp operations increasingly relied on the efficient exchange of population data within the government. Identity cards were frequently sent to a central processing institute in Berlin, tasked with managing concentration camp labour with processes universally standardised. (35 p. 428) In 1944, Himmler established the *Mechanised Central Institute for Optimal Human Registration and Evaluation* which improved the efficiency of extermination through centralisation and automation.[86] (4 p. 33)

The government's tight grip on identity management was pivotal in facilitating the era's heinous crimes. While not immediately obvious to the casual observer, the link was painfully evident to those who experienced it, including my *Oma*. A heavy atmosphere of defeat replaced the earlier zeal of the Nuremberg Rallies, as people, with slumped shoulders and averted gazes, moved through the streets, acutely aware of the oppressive web that ensnared them. The government's detailed knowledge enabled precise control over everyone's actions and dictated their societal involvement.

Conditional Access to Society

In societies where identity determines existence, conditional access acts as an invisible barrier that defines life's limits. In a conditional-access society, citizens must prove their legitimacy to the government, rather than the other way around. In Nazi Germany, this societal key was made of thin, folded cardboard and bore the name, *Kennkarte*.[87]

Requesting Permission to Exist

The *Kennkarte* system formalised the end of an evil metamorphosis. Initially sold to find enemies and access benefits, identity

---

[86] This institute was known as the *Maschinelle Zentralinstitut für optimale Menschenerfassung und Auswertung.*

[87] *Kennen* is German for knowing a fact or person and *Karte* means card or descriptive document. The *Kennkarte* identity document was adorned with the *Parteiadler* [Nazi party eagle] and citizens were required to carry the document at all times.

registration had become the tenuous condition for existence. Established by the 1938 *Passport Decree*, the *Kennkarte* was an identity card that granted the government arbitrary control over individual rights and status. (23 p. 69) It enabled the immediate denial of goods and services based on identity markers, most notably the red 'J' identifying Jews, allowing widespread enforcement by officials and civilians alike (Figure 17). So began the era of legislation-based trust, and in the words of Götz Aly and Karl Heinz:

> ... *a graded system of rewards and punishment*
> *for selection and eradication. (4 p. 2)*

The *Kennkarte*, which had to be carried at all times, guarded every transaction. Some documents unlocked food and freedom, while others did not. Those holding a *Kennkarte* stamped with the dreaded 'J' were subjected to higher taxes and lower wages for the same work, all without labour benefits. Their children were even denied the basic joy of buying candy.

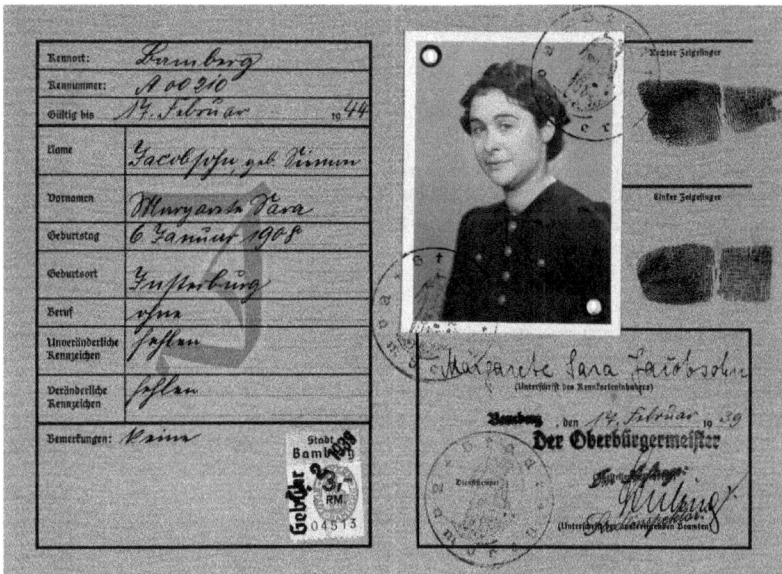

Figure 17. *(Deutsches Reich Kennkarte) issued to Margarete Sara Jacobsohn dated February 17, 1939, and stamped with a red letter 'J' for Jude [Jew]. Fingerprints visible. (65)*

Identification became the gatekeeper in all aspects of life. It was mandatory for all government interactions, including police stops, passport applications, and obtaining marriage certificates.[88] (4 p. 51) The same year the *Kennkarte* was introduced, the government restricted economic opportunities by using identity traits to bar people with certain biological attributes from many professions and corporate governance roles, including board positions in Jewish organisations.[89, 90] (23 p. 67) Additionally, professional titles, like doctors, now required proof of Aryan identity. (4 p. 61)

The outbreak of war in 1939 led to curfews and restricted shopping times for Jews.[91] The rationing system limited their food choices, enforced through mandatory presentation of a *Kennkarte* to redeem food coupons. By 1940, Jews were barred from (buying or having) certain foods, while other shoppers skipped queues during their allocated shopping time.[92] Survival often relied on circumventing the identity system through connections with sympathetic vendors.[93]

Once civilised people watched as selected groups of men, women, and children starved in the streets, thankful their government had

---

[88]　In a similar way, expansion of modern digital registration requirements can be easily made once the infrastructure is in place.

[89]　This definition is important. Being categorised as Jewish in Nazi Germany did not necessarily mean an individual was of the Jewish faith. It could simply mean that an ancestor had at one time been a Jew, sometimes unbeknownst to the individual being classified. As such, renouncing faith was no escape because the term was considered to reference a biological attribute, offering a warning about the risks of biometrics.

[90]　For Jews, the list of vocational prohibitions was lengthy. It included the auction, security, real estate (including commercial brokerage for real estate contracts and loans), commercial marriage brokerage, tourist guide, nursing, legal, patent attorney, and notary professions. (23 p. 67)

[91]　Jews in Berlin were restricted to a single hour of shopping daily from 4-5 pm, a time when many items were already sold out. The hour was often insufficient as the absence of supermarkets required separate trips to bakeries, fruit markets, and other individual vendors. Additionally, allocated shopping time often overlapped with mandatory labour hours, adding to the difficulty. (91) This hour allotment coincidentally mirrored the one-hour outdoor exercise allowance during some Australian COVID lockdowns.

[92]　Vegetables, meat, fruit, canned goods, fish, tobacco, cocoa, and milk products were conditionally prohibited.

[93]　Sympathetic vendors incurred significant personal risk by hiding food under shop counters for those the government had said should starve.

allowed them to exist for just one more day. This 'default deny' society effectively imprisoned individuals in plain sight, denying them necessities and participation in societal activities, unless they presented identification acceptable to the ruling elite. My *Oma*, despite being German, lived in constant anticipation of these interactions, where life hung on government-defined labels and ever-changing whims.

The Modern Parallels: "Screw Your Freedoms"

Is modern society witnessing a resurgence of identity-conditional access? The COVID era saw *our* once civilised culture watch, while certain groups of people were denied access to libraries, cinemas, pubs, restaurants, shops, hospitals, transport, education, and work, based on the ever-changing status of check-in and vaccination apps.

Does the red cross on a government app mirror the red 'J' on a *Kennkarte*? Recent years offered a taste of the conditional access seen in Nazi Germany. They showed a disturbing modern willingness to marginalise groups openly, based on bureaucratic disdain for their beliefs, now recorded on smartphones, rather than pieces of cardboard.

COVID era narratives convincing people that the ends justified the means were widespread, echoing similar rationalisations from Nazi Germany. Arnold Schwarzenegger's controversial remark, "Screw your freedoms," in 2021, exemplifies this. (66) Considering his Austrian background, a more nuanced approach reflecting moral strength, akin to his physical achievements as a seven-time Mr Olympia recipient, might have been expected.[94] As the Nazi regime's identity-based policies continued to 'screw' with people's freedoms, the machinery of government surveillance that underpinned these policies also grew in the shadows.

---

[94]   Austria being the victim of Germany's *Anschluss* in 1938, its annexation into the Third Reich.

Papers Please! Identity and Government Surveillance

Symbiosis of Identity Management and Surveillance

Identity registration prevents people from living unobserved by the government. This is because it operates symbiotically with ubiquitous surveillance, much like fire and air. The Nazis leveraged their extensive knowledge of the population for selective, comprehensive surveillance, which in turn reinforced identity registration compliance, bolstering the regime's control.

The cyclical relationship between registration and surveillance is well recognised. Friedrich Zahn, wartime president of the German Statistical Society, highlighted the essential connection, stating that effective inter-office collaboration hinged on detailed population identification, guaranteeing that "registered persons can be observed continually." (4 p. 105) Echoing this, in 1942, Nazi statistician, Friedrich Burgdörfer, underlined the importance of comprehensive population policies, whose effects "must be observed continuously." (4 p. 28).

The Mechanics of Control

The impact of surveillance in Nazi Germany was profound and far-reaching. Routine monitoring of telephone and telegraph communications allowed authorities to swiftly intervene in citizens' lives, unhindered by bureaucratic barriers. This omnipresent surveillance instilled a pervasive fear of arrest for any spoken word or action, creating a climate of terror that affected even regime supporters.

From the war's onset, privacy became a fleeting luxury as registration demands grew, with actions like overnight hotel stays requiring presentation of an identity card.[95] (4 p. 42) Extensive monitoring soon encompassed a wide array of personal details, from occupation and income to family information, fingerprints, signatures, and even fertility status. (4 p. 79)

---

[95] Parallels are drawn here to the QR codes and vaccine passports that governments required of people to enter all manner of establishments during the COVID era; these are hard to ignore.

Innovations in Surveillance

Surveillance also encompassed the upkeep of the population registry itself. In occupied Holland, a proposal was made to Himmler to ensure continuous registration control, by affixing a tri-layered adhesive stamp to ID cards that changed colour biannually, aiding in the regular identity updates described in Chapter Two. (4 p. 70) These cards were crucial for accessing food rations, which effectively transformed this administrative tactic into a powerful control mechanism.

Surveillance invaded the smallest details of life, extending its reach through strategic surveillance in public spaces to monitor and quash dissent. This approach is highlighted by a 1939 Nazi report on the occupied Polish population, showing the government's strategic permission of certain social activities to simplify public monitoring. The report states:

Coffee houses and restaurants, even though they were often outspoken gathering points for nationalist intellectual circles in Poland, should not be closed because surveillance seems easier here than in private meetings of conspirators that then inevitably take place...

... Because of their considerable ethnic significance, theatres, cinemas, and cabarets should only be permitted on a very small scale. The ownership of radio sets should only be possible with a special licence. An inspection of the import of gramophone records would seem to be useful.

Book production must be limited to the utmost. Newspapers and periodicals must be restricted and monitored. It is advisable to provide them with suitable material selected by German authorities. (20 p. 595)

Mobilisation for War: Identity Registration's Role

As the war progressed, the identity system expanded to include details like school performance, health and immunisation records, military service, licences, and family lineage. (4 p. 122) These aspects were quickly integrated to mobilise civilians, including the elderly and youth, for military defence roles and manufacturing. Identity registration and surveillance facilitated the conscription of men aged from 16 to 60 into the *Volkssturm* [People's Storm] on September 25, 1944. (3 p. 540) This resulted in under-equipped and un-uniformed elderly men and boys being drafted as expendable troops in the war's final months.

The Police State and Its Enforcers

Do guns, guards, and gates keep us safe? The strength of a government largely depends on its enforcement agencies. In modern Western society, there exists a common but misguided belief in the safety provided by a strong, militarised government. However, granting one entity sole control over force risks harming society. This view is rooted in the American founding fathers' understanding of government's tendency to oppress its people during times of disequilibrium, leading them to enshrine the right to bear arms in their Constitution.

The citizens of Nazi Germany had no such protections. In 1941, as my grand-uncle Manfred fell during Operation Barbarossa, the Nazi military commanded over four million men in 205 divisions. (3 p. 42) However, its internal oppression of citizens involved especially cruel forces. As domestic policies intensified, so did the use of force to implement them, relying on departments that employed terror for civilian control.

The Gestapo: Suppressing Dissent

The *Gestapo* established in 1933 by Hermann Göring, was at the core of Nazi domestic oppression, acting as the 'fact-checkers' of the time to quell dissent and instil fear.[96] Working alongside the Criminal Police, they significantly contributed to expanding identity registration, including a 1936 pilot program to register gypsies. (4 p. 3) Gypsies were an ethnic group living in Europe with origins in present-day India and Pakistan. Considered racially impure, identity registration allowed the *Gestapo* to subject these people to eugenics practices aimed at categorising them based on physical traits.

Gestapo officers managed the *Volkskartei* identity cards, and enforced discriminatory practices with harsh penalties, such as searches in Jewish homes for black-market food.[97] In a foreboding 1941 speech, *Gestapo* chief, Reinhard Heydrich, stressed the need for:

> ... *an inventory of the people that once and for all registers all the persons of the Protectorate and sorts them according to certain aspects. (4 p. 52)*

By July 31, Heydrich and his *Gestapo* spearheaded the mass deportation of Jews from Europe, with his concept of 'certain aspects' extending beyond Jewish identity.

The Sicherheitsdienst: Intelligence and Surveillance

The *Sicherheitsdienst* (SD), served as the Nazi regime's clandestine eyes and ears, acting as its intelligence agency.[98] They implemented pervasive surveillance. No place—not even churches—escaped its observation. (15 p. 57) This relentless monitoring even extended within the Party, to help guarantee Nazi officials' absolute loyalty to

---

[96]   *Gestapo* is an abbreviation of *Geheime Staatspolizei* [Secret State Police].

[97]   Food prohibited to only Jews, with violators facing severe penalties, like fines and confiscation of property.

[98]   Security Service.

Hitler. This omnipresent surveillance curtailed freedoms and played a key role in eroding citizens' rights.

The Schutzstaffel: A Multifaceted Force of Terror

Who monitors the monitors? Embedded in the Nazi control apparatus was the *Schutzstaffel* (*SS*), or Protection Squad, initially supervising the *Gestapo* and *SD*. The *SS* juggernaut, known for its multifaceted roles from policy enforcement and military operations, to running death camps, epitomised brutality, notably in the massacre of *Oradour-sur-Glane*, France. Far beyond instilling terror, the *SS* played a major role in destroying evidence, including incriminating documents and identity records associated with concentration and death camps, ahead of liberation. (3 p. 38)

Like the *SD*, the *SS's* ruthlessness was not limited to external enemies. It also murdered members of the regime's own paramilitary Storm Division, known as the *Sturmabteilung* or *SA*, deemed inadequately loyal. Feeding on its own, the cold and calculated actions of the *SS* deeply tarnished the pages of history with indelible black marks.

Einsatzgruppen: The Mobile Killing Units

Special Action Deployment Groups [*Einsatzgruppen*] were notorious subsidiaries of the *SD* and the Security Police, colloquially known as mobile killing units. Like census takers, *Einsatzgruppen* troops shadowed the army into newly-captured territories, with a clear mission: to identify and eliminate any form of resistance, targeting victims by age, gender, occupation, and religion. In only the first three months of the war, the *Einsatzgruppen* murdered 50,000 Poles, only 7,000 of whom were Jews. (67) Their notoriety for mass murder was infamous, exemplified by the staggering 33,771 Jews slain in only two days near Kiev in September 1941. (25)

Simply stated, personal identification in Poland was a matter of life and death under Nazi rule. Strict adherence to registration was enforced; failure to register or falsification of documents, often resulted in deportation to concentration camps. (4 p. 43) Not having an ID card could result in execution. (4 p. 5) The *Einsatzgruppen*, tasked with eliminating resistance, targeted a wide range of individuals, from boy scouts and medical professionals to lawyers and Catholic priests. Even the future Pope John Paul II had to work in a quarry to avoid being sent to Germany. (68) These actions revealed the broader aim of the Nazi war effort: to systematically eliminate anyone based on various identity factors, as recorded in detailed population records.

Chapter Takeaways

How can you protect and value your identity in a world increasingly inclined towards digital government identification? Considering the historical and contemporary challenges associated with government-managed identity systems, it is imperative for people to take proactive steps to legally safeguard their identities and maintain their autonomy. Here are some actionable takeaways to help you protect your identity from prying governments.

In Private
Diversify Dependency

Relying heavily on government services for essential needs like childcare, healthcare, employment, and education can increase vulnerability to invasive identification practices. Diversifying your dependencies can mitigate these risks. This reduces the government's leverage over your identity and enhances your ability to opt-out of systems conditional on extensive personal information. Proactively seek private or community-based alternatives in these key areas to strengthen your personal resilience and reduce vulnerability by not "putting all your eggs in one basket."

In Public

Engage in Policy Advocacy

Actively support lawmakers who promote limited government involvement in identity management. Participate in public hearings, sign petitions, and communicate with your representatives to express your concerns about centralised digital ID.

People often vote for political parties out of habit, like backing a sports team, even when those parties have evolved beyond their original politicians and policies. Flexibility is key; being open to 'switch sides' allows us to hold politicians accountable by endorsing or penalising their policies based on merit. Voting for smaller, issue-focused parties can be a good strategy. These parties are often more attuned to grassroots concerns and less dominated by career politicians, making them more responsive to public opinion due to their dependence on every vote.

The recent review of the Digital ID Bill 2024 by an Australian Senate committee received significant negative feedback, as noted in the committee's final report. This indicates that policymakers are attentive to public opinion on this issue. By persistently voicing our concerns, we make it increasingly difficult for our perspectives to be ignored.

Prepare for Resistance

History has shown that collective action can lead to meaningful change. Be prepared to resist measures that infringe on your identity. This might mean opting out of digital ID programs or finding creative ways to navigate systems without compromising your identity. Resistance can take many forms, from legal challenges to peaceful protests. However, all require courage and conviction. The path forward is not without challenges, yet in keeping with this book's thesis, these actions are made easier when done out of love for what we wish to protect, rather than hate for what we want to avoid.

A Better Relationship with Government

Centralised identity management poses significant risks, especially under governments with unchecked authority. The Nazis used intricate institutions and legislation to disguise their true intentions. Yet, their approach to identity management, while outwardly legitimate, resulted in profound harm, highlighting the disconnect between their promises and the destructive outcomes.

Is the push by modern governments for centralised control over digital ID motivated by self-preservation rather than altruism? Presented to offer convenience and safety, such control may actually serve to entrench authority. Challenging those in power helps prevent rule-makers from disproportionately benefiting from the systems they create.

Within this narrative lies a powerful reminder of the human spirit. Never underestimate the potential for internal reform within governments, as these institutions are not monolithic. Like society at large, governments consist of individuals, many of whom are parents concerned about future generations. By supporting allies within the system, we can influence policies that prioritise individual rights in the technological landscape.

Finally, we must recognise that legal frameworks alone provide insufficient protection; we need robust technical and procedural safeguards, and absolute prohibitions that defend against the misuse of identity by any government. Our access to a free and open society depends on our collective action, and the choices we make today.

CHAPTER 6:
Modern Society

The greatest threat Australia has ever faced.

—Peter Garrett commenting on the Australia Card. (69 p. 11)

Given the horrific impact the card will have on Australia,
its defeat would almost be worth fighting a civil war for.

—Dr Bruce Shepherd, AMA (Australian Medical Association) president,
on the Australia Card proposal 1987. (69 p. 12)

What is at stake is not just catching a few tax avoiders.
It is not even the efficiency of policing. It is not the defence of
innocent and law-abiding citizens from law breakers.

What is at stake is nothing less than the nature of our
society and the power and authority of the state in
relation to the individual.

—His Honour Justice Michael Kirby, President of the New South Wales Court of Appeal,
and former Chairman of the Australian Law Reform Commission,
on the Australia Card proposal 1987. (12 p. 180)

I fear what can go wrong. I shouldn't need to have this fear
as a citizen.

—Personal contact and technology professional from India, requesting to
remain anonymous, commenting on Aadhaar, India's centralised government
digital ID system 2024.

A New Life, a New Beginning, and an Old Adversary

Starting a family brought great joy to my grandparents during the 1950s. They cherished my father's early years, watching him grow, play, and attend school in the relative safety of post-war Germany. Yet the indelible scars of Europe's horrific identity abuse never truly healed. Eventually, they decided to resettle as far from the nightmares as possible—Australia.

A country that had declared war on my *Oma's* homeland two days after she had watched Nazi tanks roll into Poland, was now welcoming a new generation born from that very conflict.[99] (3 p. 528). Having experienced the tyranny of human commodification, the family sought a home with certain qualities. Australia stood out with its temperate climate and educated, compassionate people. But above all, they admired its fierce sense of community, and history of resistance against government control and coercion.

In this chapter, we conclude my *Oma's* journey, as she fought an unwelcome resurgence of government identification in her adopted country. By scrutinising the triumph over the Australia Card, and the ongoing battle with India's Aadhaar, the world's largest national biometric ID system today, we highlight the universal risks these systems present in modern society. Most importantly, we explore the merits of living in a country woven from threads of citizen vigilance, and self-reliance.

Forged in Colonial Struggle: The Birth of Australian Identity

How had Australia's people developed their strength and independence? As early as 1854, in the gold rush of colonial Victoria, this spirit emerged among hardworking prospectors. Family men, reformed convicts, and migrants alike faced increasingly oppressive mining licenses, which activist Geoffrey Blainey described as:

---

[99] Along with Britain, France, and New Zealand, on the same date.

... the identity cards of the time. (2 p. 272)

Peaceful attempts to resolve these issues failed, with miners driven to breaking point by colonial forces conducting brutal tent-by-tent licence hunts. Bonded by a shared suffering, a group of miners voluntarily organised an armed rebellion behind a stockade at the Eureka goldfield in Victoria, Australia (Eureka Stockade).

On December 3, government troops attacked, killing at least twenty-two miners. Yet the rebellion was considered a success, heralding political change that transcended mere licenses. The victory over the government-issued card symbolised a pivotal moment for individual identity in the young nation of Australia. It was a direct statement that firmly denounced people's livelihoods being contingent on government approval.

Confirmed on the Battlefield: Community Solutions

The same spirit was showcased on April 25, 1915, as Australia was coming of age. During the Gallipoli campaign in the First World War, 16,000 Australian and New Zealand Army Corps (ANZAC) soldiers, many of them boys on their first trip from home, bravely faced entrenched Turkish machine gun fire when landing on the rugged cliffs of the Gallipoli peninsula. By the end of the day, 2,000 ANZACs had fallen with thousands more to perish in dire conditions over the next eight months. (70)

However, it was camaraderie, not complaint, that shaped the Australian experience. The extreme hardships fostered loyalty, with men driven by relational trust and self-sacrifice, beyond mere orders. Private John Simpson, a stretcher-bearer with the 3rd Field Ambulance Brigade, became legendary for these qualities.[100] He tirelessly transported wounded soldiers to beachside medical stations at ANZAC Cove under relentless enemy fire using a donkey, bravely

---

[100] Formally John Simpson Kirkpatrick.

saving more than three hundred men over twenty-four days until his untimely death. (71)

Voluntary Adoption of Responsibility: Cause by Choice

Why did these moments in history capture my family's eye? The Eureka miners demonstrated how even small communities can voluntarily produce large changes by supporting common values. Equally, Australia's engagement in the First World War, showcased the potential of a nation committed to a cause by choice. With the country refraining from conscription, soldiers like Simpson had all willingly enlisted.[101] Then Prime Minister Billy Hughes had entrusted the people with this choice through two referenda in 1916 and 1917. Mature public debate saw communities twice reject the use of government power, to send men to their deaths.

Impressively, even the soldiers in the trenches opposed conscription. Despite its potential to help, they were unwilling to force their horrors onto others. Back home, another army of sorts played a pivotal role in maintaining the voluntary nature of the war effort. With Australia being one of the few nations allowing women to vote, wives, daughters, and mothers alike leveraged their positions to sway the referenda results.

Community Driven Solutions

Despite this vote, the nation was far from shying away from the fight. Boys as young as fourteen tried to enlist by falsifying their ages, and recruiting sergeants often allowed shorter statured men to meet height requirements by standing on their toes. Entire towns saw their working-age men leave to enlist, motivated by duty to their country-

---

[101] Interestingly, British officials had attempted to pressure Australia into implementing conscription to facilitate a stronger, albeit unwilling, military presence in the war. Regardless, the people were given the ultimate decision.

men when hearing of the events at Gallipoli. These self-motivated acts could not have contrasted more with the government roundups of unwilling conscripts my family had seen in Germany.

Taking conscription off the table had birthed the community-driven idea of recruitment marches. Spearheaded by 'Captain Bill'—a plumber named William T Hitchen—a group of rural men undertook a 320-mile march to Sydney, in all weather conditions, to bolster enlistments.

The sight of these men sparked enthusiasm in the towns they marched through, swelling their numbers to 300 along the way. Despite never returning from deployment, Hitchen's grassroots initiative demonstrated the power of local action, without government intervention.[102] His march inspired others nationwide, providing vital support to deployed troops.

The Ethos of Self Resilience

This spirit of voluntary self-sacrifice became central to Australian identity. The country exemplified the power of free societies in committing to collective values, contrasting with the acts of self-preservation often seen under coercion. So long as the government kept out of the way and people were able to try their hand at a livelihood through hard work, people were willing to support the culture that supported them.

The ethos of loyalty, and Australia's well-known concept of a 'fair go' emerged from this history. Isolated on a vast continent, Australians realised the importance of self-reliance and personal relationships in overcoming adversity, and tyranny. They learned that contributing solutions yielded better results than relying on government,

---

[102] The State Recruiting Board and the Defence Department had denied support, claiming recruitment marches would be expensive, cumbersome, and ineffective.

with heartfelt loyalty to those nearby being more vital than arbitrary adherence to distant authorities.

Paying it Forward

This steadfast defence of identity and self-responsibility, from the individual to the national level, was what drew my family to Australia. They had seen firsthand the repercussions of relinquishing personal sovereignty to the government. Now, they were proud to contribute to a nation that espoused a different mindset. Cementing the move, my father married my mother, an Australian lady from a family of naval heritage.[103] I became the first in his lineage to be born on Australian soil.

I was blessed to be raised in a family that conveyed these profound lessons (Figure 18). My childhood brims with fond memories of

Figure 18. Me with my mother and uniformly dressed great-grandparents. From the Battle of Passchendaele to Australia. I was the first of my father's lineage born in Australia.

---

[103] My maternal grandfather had been posted to the light cruiser *HMAS Sydney II* when it was sunk by the German auxiliary cruiser *Kormoran* off the Western Australian coast in 1941. By a twist of fate, he missed the ship's deployment, a serendipitous turn that not only saved his life but also made possible the birth of my mother.

education and play. My *Oma* would share stories, take me swimming, and teach me Origami, while my grandfather introduced me to woodworking and piano.[104] They modelled love and independence in their interactions, educating me about identity and trust, without me even knowing it.

Yet, as I enjoyed my childhood, a sinister force was taking root. It was the same evil that had grown as Nadine Schatz cycled the streets of Paris in her youth. Now, it was reemerging halfway across the globe, in a country on the cusp of its bicentenary. This time, it bore the name ... *The Australia Card.*

A Win: The Australia Card

In the mid-1980s, Australians were wrestling with a growing economic crisis, and frustration over perceived inequality. The Australia Card, a national identification scheme, emerged as the government's panacea.[105] On one hand, it promised bureaucratic efficiency, improved healthcare, and crime prevention. On the other, it sought the eradication of tax fraud, welfare abuse, and illegal immigration. (12 p. 176) Yet its true purpose was clear: to link life's essentials, like employment and access to property, to a mandatory government ID card.

Identity

To achieve this, the federal government was to assume control of identity management. Citizens would then need to show an ID card for all manner of previously unhindered activities. Issues in one

---

[104] I recall one incident at a public pool when I was able to rescue a child who I had seen floating face down. Being only a boy myself, I had struggled to bring the child to the edge where his unaware mother had frantically taken him. It was my first experience with the real effects of community help. When seconds count, government assistance is minutes away. My *Oma* bought me an ice-cream as a reward when leaving the pool that day.

[105] The Australia Card was a major proposal made in 1985. It first appeared in a government whitepaper.

area of life could now spill over into others.[106] By tracking this data, authorities could act on individual grievances rapidly, turning each reportable event into a high-stakes game of cat and mouse.[107] Simply making an offer to purchase property, for example, could see the police arrive at the real estate agency for an unrelated welfare issue.

Schedule One of the Australia Card bill listed the identity traits to be centralised under the scheme. These included nicknames, travel plans, genealogical data, and passport details.[108, 109, 110] Employment records, citizenship status, and Medicare eligibility were also fair game, along with a two-year residential history and a digitised image of everyone's signature.[111]

My grandparents heard of the plan on a morning radio broadcast. While sipping their coffee, the announcer sold the Australia Card as a tool for national benefit. Yet, having lived through it before, my *Oma* instantly recognised the shift of basic rights into privileges, granted to citizens like treats for obedient dogs. Horror-struck, she faced being photographed and catalogued for the second time in her life, and me for the first.

That sobering news bulletin stirred dormant traumas in my *Oma*. It echoed the "surreptitious social engineering" that Joseph Göbbels had unleashed over the radio waves in a different era and language.[112] (12 p. 176) The news left her stone cold, like her now untouched

---

[106] Previously, individuals were identified differently between government departments. A Tax File Number (TFN), for example, was meaningless outside the Australian Taxation Office (ATO). The Australia Card sought to dismantle this safeguard by allowing isolated issues to affect all other areas of life.

[107] Authorities could act as quickly as the technology of the 1980s allowed. Today, authorities can act in real-time.

[108] Visitors needed to provide arrival dates, entry status, planned and actual departure dates, entry permit expiry dates, and more.

[109] This even included parental names and birthplaces.

[110] This also included passports issued by other countries.

[111] The actual wording in the bill illustrated the de-humanisation of the plan: "A digitised image of the specimen signature." (72 p. 135)

[112] Liberal Senator for New South Wales, Christopher Puplick, used this phrase as he called out the Australia Card proposal for what it was.

coffee. The predator she had once narrowly escaped all those years ago, had found her again.

Trust

One of the first similarities with the *Volkskartei* that my *Oma* noticed, was the treatment of those with opposing views. Back then, they were branded 'anti-social' and 'useless'. Now they were being labelled 'friends of tax-cheats'. (69 p. 11)

Shallow justifications were the next telltale sign of nefarious legislation. In her youth, abdication of trust to the government was sold as "the protection of the people against criminals." (4 p. 38) In adulthood, it was now being labelled as a move "towards fairness and equity."[113] (12 p. 12) The absence of precise definitions, and fair debate, was immediately evident to her, and she was not alone.

Beneath the façade, many saw the Australia Card as a vehicle to erode rights for the unregistered, rather than the unlawful. The government even conceded the card's mandatory nature was at odds with personal liberty. (12 p. 26) The term 'pseudo-voluntary' was coined to navigate this issue, despite the government admitting that those without cards would have:

> ... *great difficulty in surviving. (12 p. 59)*

The Disproportionate Price of Non-Compliance

My *Oma's* sense of *déjà vu* heightened when reading the punitive language that permeated the Australia Card bill. Part Eight was solely devoted to offenses. The terms 'imprisonment' and 'penalty' appeared a staggering twenty-eight and eighty-four times, respectively. Forty-eight instances cited fines of AUD $20,000 (equivalent to AUD $63,096 in 2024 terms).[114]

---

[113] In February 1986, the Government produced a 300-page submission entitled, *Towards Fairness and Equity*. (12 p. 12)

[114] This is based on the consumer price index (CPI) measure of inflation. (103)

Failure to attend a government interrogation program, known as a 'compulsory conference', attracted an AUD $1,000 fine (now AUD $3,155) and six months in jail. (72 p. 113) Individuals unable to explain a damaged card, faced an AUD $5,000 penalty (now AUD $15,774) and two years in prison. Even minor lapses, like not reporting a lost card within twenty-one days, incurred an AUD $500 fine (now AUD $1,577). (72 p. 128) Page after page read similarly, leaving my *Oma* near tears.

The experience had immediately transported her back to the random police checks of her youth. She read the proposed legislation with the same eyes that had seen the results of similar policies unfold in the streets of Nazi Germany. She now faced a government wanting to impose an AUD $20,000 (now AUD $63,096) fine for failing to produce a card when demanded by authorities. (69 p. 13) Yet paradoxically, the same government was threatening two years' imprisonment for any member of the public making the same demand of others.

Even worse, these new offenses were linked to vague definitions. Using an Australia Card "with intent to deceive" or possessing a document that could be "mistaken for" an Australia Card were heavily punished. (72 p. 122) Yet proving intent or deception in a court of law is challenging. Conclusions are subjective and vary between juries.

A Shift in Trust: The Presumption of Guilt

My *Oma* understood that a government forcefully involving itself in identity management, is not merely coercing behaviour. It is cultivating a perception that those engaged in reportable activities, albeit legally, are inherently suspicious. This viewpoint was not unique. Many, including then president of the Australian Medical Association, Dr Bruce Shepherd, publicly cautioned that the Australia Card would:

... turn Australian against Australian. (69 p. 12)

The presumption of innocence until proven guilty is a critical component of a civilised society. A culture that presumes innocence views a neighbour's new car with appreciation and positivity, trusting the assumption that they did well to earn it. The inverse is to assume that a neighbour's new car was acquired via dubious means. By defaulting to suspicion, any positive impression of a neighbour must first be proven.

The more a society abandons trust as a commodity earned within communities, the more centralised identity will need to be "produced on demand for purposes beyond the government's intention." (12 p. 178). For me, the concern is that *today's abnormal will become tomorrow's normal.* Interactions that now start with a handshake become an exchange of identity documentation as centralised identity credentials, analogue or digital, become increasingly "produced as a normal practice." (12 p. 178)

Perhaps the most eloquent expression of this sentiment came at the time from His Grace, the Right Reverend Michael Challen. He commented on the Australia Card in his capacity as Bishop of the Anglican Diocese of Perth, and Chairman of its Social Responsibilities Commission:

No family, no community and no nation can really work very happily except on the basis of trust, and trust is not a commodity which one gives to another.

Trust is a quality and a response which you evoke out of another by, in fact, entrusting yourself to that person or group. (12 p. 178).

Manufactured Justifications and Deceit

With public dissent growing, a Joint Select Committee was convened on December 4, 1985, to dissect the Australia Card proposal. Under this scrutiny, the government's justifications fell like a trail of dominoes. What became clear was not that the government had seen a problem and wished to implement centralised identity registration as a solution. Rather, the government had seen an opportunity to implement centralised registration and had manufactured claims and figures to justify their position.

Australian Taxation Office (ATO)

To gauge the claimed economic benefits, the committee interrogated the Australian Taxation Office (ATO). The ATO acknowledged that hidden commercial activity, known as the 'black economy', was immeasurable. However, they simultaneously quantified the resulting tax evasion at AUD $3 billion. (12 p. 47) Moreover, the department claimed that seventy-five percent of taxable income could be identified with the Australia Card. Yet further probing revealed this seemingly precise calculation was actually nothing more than "a qualitative assessment of how tax evaders operate."[115] (12 p. 49)

Department of Social Services (DSS)

The arguments for combating social security fraud also unravelled. The Department of Social Services (DSS) candidly acknowledged that false identities did not notably impede their operations.[116] They further confirmed that a national identification system would yield "no net gain." (12 p. 52)

Paradoxically, the department still crafted a cost-benefit summary that conveniently concluded the costs of implementing the program, were perfectly balanced by its earnings.[117] (12 p. 52) Further scrutiny showed that the DSS had tailored these figures to back the government's push for the Australia Card. Notably, the Secretariat, when drafting the government submission, aligned the projected benefits

---

[115] A qualitative assessment is based on a subjective and relative guess of what is better or worse rather than facts that can be measured.

[116] This was the precursor to what is now Services Australia.

[117] Specifically, this equated to AUD $18.2 million over ten years. (12 p. 52)

with the proposal's annual costs. (12 p. 53) In other words, they had made them up!

Department of Immigration and Ethnic Affairs (DIEA)

The Department of Immigration and Ethnic Affairs (DIEA) also found itself in an awkward position. The claim that the Australia Card would identify 60,000 jobs filled by illegal immigrants was revealed as mere guesswork. (12 p. 57) That figure had been based on an unpublished departmental study conducted years earlier, using a flawed methodology. The figure was revealed under questioning as having come from:

> ... a crystal ball type of calculation. (12 p. 58).

Health Insurance Commission (HIC)

Yet perhaps one of the most damning revelations came from the Health Insurance Commission (HIC), the agency that would have been tasked with administering the scheme. In a report dated February 26, 1986, the HIC craftily suggested mitigating public backlash by rolling out the system in phases. They proposed starting with the centralisation of less sensitive data, gradually incorporating more, so as not to raise public attention. (69 p. 14)

Technology

While many of the Australia Card's justifications were fluid or fabricated, its technical infrastructure was to be concrete. It would connect the federal government with state government databases.[118] Further, new processes would enforce the currency of data exchanged. Just as in Germany, local title registrars, for example, would be obligated to provide the ATO with ongoing sales updates. (12 p. 28) Yet the proposal called for not only a broadening of data collection but also of data access.

---

[118] Notably those of Births, Deaths, and Marriages (BDM), and land title registries.

Computer Data Matching

The government initially wanted to share identity data across federal departments that included Defence, Education, Veterans' Affairs, Foreign Affairs, and Immigration & Ethnic Affairs. (36) However, the bill's wording allowed for permanent, two-way data sharing more broadly. This made expansion a trivial legislative exercise once the technology was in place.[119]

The Joint Select Committee identified the wholistic processing of this combined data as the technical equivalent of a warrantless search. Such activity is inherently not based on suspicion of wrong-doing, but on the presumption of guilt.[120] A warrant is a targeted tool applicable when evidence precedes suspicion. Using computers to match generic identity numbers between systems, reverses that order by casting:

> ... whole categories of persons under suspicion. (12 p. 140)

The Committee additionally recognised that identity data matching was a powerful location tracking and surveillance tool. While the government did not widely publicise it, they acknowledged that the registry could pinpoint "the current whereabouts of a person." (12 p. 138) This admission came despite the Australian Federal Police (AFP) stating that they were already capable of finding people, without the assistance of the Australia Card. (12 p. 139)

---

[119] Today, fifteen government services are linked in one place on the MyGov site. These include the Department of Veterans' Affairs, the Australian Taxation Office, Medicare, and the National Disability Insurance Scheme (NDIS). (93)

[120] Such real-time warrantless policing is fundamental to the COVID era check-in apps which populations have accepted without consideration of potential implications. There is significant power inherent in apps that implement real-time warrantless surveillance. Hence, it could be argued that their rapid rollout was a desired outcome rather than an ill-considered side-effect, regardless of the claimed motivations of the time. The immediacy of such policing is restricted more by the technology of the time than any legislative means.

Who Polices the Technology?

Access to the identity data was to extend far beyond the departmental level. Like the Byzantine Command Centre from Chapter Three, government departments are, after all, merely abstract collections of individuals. Around 50,000 bureaucrats were to be permitted to use the registry. (12 p. 197) Each of these was a potential source of leaks, being susceptible to bribery, blackmail, and malfeasance. The Australia Card was to provide a one-stop-shop for identity information, potentially benefiting disgruntled lovers, and domestic abusers alike.

Accordingly, the expanded access was identified as a health and safety risk. For example, linking current and previous addresses and names dissolves protective measures designed for those escaping violent relationships, or to hide trial witnesses. Under such systems, those legitimately seeking to disconnect from their past, find themselves perpetually haunted by it instead. Like unhindered vigilantes, shadowy figures at government terminals could act on rogue retribution, rather than simply taxation tracing.

Yet perhaps most important among the many revelations about technology's impact on society came from Mr Frank Costigan, QC. The former Royal Commissioner stated:

You ultimately have to prove that the computer is wrong and you can just imagine the problems if something has gone wrong and you have to persuade the person across the counter that you are right and the computer is wrong.

I think it really is a big change in the way in which we have lived in our society. If you introduce something like a national identity card – again, going down the track 10 or 20 years, seeing it as it would be then – I think you really have changed the kind of society we have. You have got to be pretty satisfied that

> *the benefits you are getting out of that, justify that.*
> *I certainly am not satisfied. (12 p. 179)*

Computerised judgments foster binary thinking. Despite addressing contemporary issues, citizens were mindful of the system's potential to reshape society in the coming decades. Awareness of the era's technology, marked by the recent introduction, and scepticism, towards credit cards, influenced perceptions not only of present but also of future governmental applications.

Government

De-humanisation

For these reasons, the Australia Card aimed to be the most formidable data surveillance and social control system the nation had seen. From its inception, the proposal embodied the Personage Principle doctrine, born from Erwin Cuntz. Far from hyperbole, its goal was that of any universal identity system, to regulate access to resources through simplistic human categorisation.

De-humanisation was evident in bill's very wording. In it, people were reduced to the term *'card-subjects'*. There were disturbing parallels with the *Kennkarte*. The card itself was to be made visibly distinct—different colours were suggested—based on characteristics that held no social or legal bearing. (12 p. 142) Doing so, would visibly leak irrelevant identity data in many daily interactions, making it an ideal platform for all sorts of community-imposed discrimination.[121]

Conditional Access to Society

From their desks, bureaucrats would hold the power to deny strangers access to tertiary education, government benefits, and even home loans. (12 p. 216) Politicians could legislate higher taxes, prohibit

---

[121] Including residency, citizenship, and Medicare entitlement status.

investments, and even ban the use of safety deposit boxes for those who failed to register their identity.[122, 123] In healthcare settings, the sick, injured, and elderly faced treatment based on their card's legal status, rather than their medical needs. To exacerbate the situation, customs officers at airports could deny departure, to anyone seeking to escape the turmoil.[124]

Despite being initially presented as "a record of entitlement to Commonwealth benefits," the Australia Card proposal was revealed under scrutiny as the beginning of "a record of those identities that are entitled to operate in the Australian community." (12 p. 177) Whether through malice or ignorance, it was, as Professor Geoffrey Walker articulated, a shift "away from merely entitlement to government benefits, to an entitlement to exist." (12 p. 177) He further stated:

> *The turning of Australian citizens into 'entities entitled to operate' is symptomatic of the whole approach of the Australia Card. (12 p. 177)*

Once more, an engaged population used their collective power to critically scrutinise their government's proposals, recognising the early indicators of de-humanisation and potential control over property within their culture—a culture they were guardians of. They understood the importance of safeguarding individual property and labour rights for future generations, including those, like me, still too young to protect these rights themselves.

Property and Labour Control

Punishments extended beyond fines and imprisonment, introducing severe property controls that echoed the *1938 Ordinance on the Use*

---

[122] As a punitive measure, an elevated 49 percent income tax rate was proposed for cardless subjects just as higher tax rates applied to Jews in Nazi Germany.

[123] This was applicable whether it was interest bearing or not.

[124] Passports were not to be issued to cardless individuals. Noteworthy is the fact that Jews in Nazi Germany were pressured to emigrate whereas Australians were, conversely, to be held captive domestically under the proposed system.

of Jewish Assets. If the government denied a card to an individual, the victim would be barred from accessing existing bank deposits, opening new accounts, conducting financial transactions with solicitors, receiving funds in trust, and sending funds abroad. (73) Furthermore, the cardless were to be denied access to interest-bearing investments, including property trusts and shares, with broad prohibitions also placed on investment transactions, rental income, and other non-wage income.

Control of real estate owned by cardless subjects was also revokable under the proposal, with the purchase or rental of houses or land attracting an AUD $5,000 (now AUD $15,774) fine. (73 p. 13) These measures, in essence, legalised state confiscation of private property control without compensation, like Nazi legislation had half a century earlier. The bill's reach even extended to labour control, making employment contingent upon identity registration. It prohibited employers from hiring or paying cardless workers, mirroring the 1935 *Arbeitsbuch*. This left victims with a difficult choice: work without pay or resign. (73)

Complicit Private Businesses

Just as the Nazis had befriended private industry, the Australia Card was to leverage non-government bodies under the friendly sounding 'companion entity' system. This arrangement forced businesses, societies, trusts, partnerships, and clubs, under threat of penalty, to collect the Australia Card number for those transacting with them. (12 pp. 199 - 200) This effectively deputised many institutions as so-called 'enforcing organisations', expanding government surveillance beyond its native capacity. (36)

Initially, these organisations were only to monitor areas related to taxation, welfare, and immigration. However, because the reporting obligations could be specified separately from the bill, there was ample scope to expand the applications without amending the Act. (36)

The Pushback

In time, a broad spectrum of Australians understood the implications of the Australia Card, spawning a strong and ever-growing grassroots opposition. The collective evidence from the ATO, DSS, and others, revealed that the proposal was built on unsubstantiated, even deceptive grounds. The fusion of identity, technology, and punishment was particularly egregious to the public. These contradicted the values that the Eureka rebels had fought for, and the ANZACs had later defended.

Public Protests

The nation's revulsion towards the Australia Card was palpable and unprecedented. Widespread public resistance was spearheaded by diverse groups, including politicians, authors, journalists, business leaders, farmers, doctors, barristers, sports stars, musicians, and academics. Individuals took to newspapers, frustrated at being reduced to numbers. Cartoonists fought with pencils, portraying then Prime Minister Bob Hawke in a Nazi uniform. (69 p. 12) Masses of people rallied in the streets, also displaying Nazi uniforms in disgust.[125]

Public resistance was surpassed only recently by the mass freedom protests of the COVID era. Much like the Eureka rebels, informed and concerned individuals found unity in opposing a shared adversary, this time without violence. Their resistance was so influential that it led to the establishment of a trust, which later evolved into the Australian Privacy Foundation, still active today. (74)

---

[125] A notable instance, held in Perth on September 23, 1987, drew an impressive crowd of up to 40,000.

Parliamentary Protests

Objection extended far beyond the general public. In a speech to the House of Representatives, Mr Lewis Kent MP argued that the Australia Card should be more fittingly called the *'Hitlercard'* given its authoritarian nature. (12 p. 181) He evocatively referenced personal friends who bore numbers not printed on plastic cards but tattooed on their forearms.[126] Through Mr Kent's voice, the parliament resonated with chilling cautionary tales delivered from Holocaust survivors. These warnings firmly linked the Nazi regime's use of *Volkskartei* and *Dehomag* identity cards, with the atrocities of that era. (12 p. 181)

Members of Parliament tend not to make public comparisons to Nazi concentration camps flippantly. Yet these comparisons were justified and reflected the nation's outrage. Everyday people made the issue their business, with the Joint Select Committee mailing over two tonnes of material in response to public requests. (12 p. 12)

The Result

As much as the Australia Card met its demise through its obvious flaws, it was also an idea not fit-for-purpose. Although the government introduced legislation in 1986, the overwhelming backlash led to the bill being blocked twice in the Senate by the opposition and minor parties. This ultimately triggered a double dissolution, a parliamentary procedure to resolve deadlocks between the lower and upper houses, resulting in a national general election in 1987.

The ability of the Australia Card to impact everyone's humanity saw the issue cut through all divisions of society, and it was the absence of acquiescence and apathy that delivered my generation a different fate to Nadine's. Yet, not all countries have been as lucky as Australia when it comes to overcoming modern attempts at imposing

---

[126] The arms of Nazi prisoners were crudely tattooed with identification numbers in concentration camps.

centralised identity systems. It is time to turn our attention to just such an example.

A Loss: Aadhaar
Context and Motivations

Reflecting on the victory against the Australia Card highlights that the fight to protect personal identity from government overreach is timeless and universal. The relevance of these lessons is under-scored by the challenges presented by Aadhaar, India's centralised digital ID system. Aadhaar raises similar concerns about privacy that once mobilised Australians, shedding light on the broader risks and implications of digital ID.

Indians, like many global citizens, are becoming increasingly entan-gled in a complex digital web, with the Aadhaar system serving as the central spider. It was initially met with public opposition and concerns over potential abuse that persist today. Despite this, the government has largely overlooked these objections, leading to ongoing issues of distress and disenfranchisement among citizens.

Rajiv Gandhi and the Promises of Integrity and Efficiency

Aadhaar's integration into India's intricate social and class systems is a vast subject, potentially spanning an entire book. However, this section aims to underscore the contemporary relevance of the themes we have discussed in this book, aptly introduced by the notable 1985 quote from former Indian Prime Minister Rajiv Gandhi:

> *Of every rupee spent by the government,*
> *only 15 paise reached the intended beneficiary. (75)*

Aadhaar, we're told, was designed to improve welfare distribution by directly delivering government aid to beneficiaries, thereby reduc-ing leakage and corruption. In 2017, the Supreme Court of India hailed Aadhaar as a potential remedy to this enduring 'malaise'. (75)

An Anonymous Insider

In researching Aadhaar's complexities for this book, I sought first-hand insights beyond academic sources to capture the essence of living with a centralised digital ID system today. Fortunately, I have collaborated with numerous Indian technology professionals throughout my career. One, now living in Australia, offered to share his experiences under the condition of anonymity, highlighting the fear of repercussions for critiquing the system—even abroad.[127] We'll refer to him as Ishaan.

His insistence on anonymity and selective communication channels underscores the deep-rooted apprehension surrounding Aadhaar discussions. Ishaan's input begins with examining the challenges of implementing such a universal system in a populous and hierarchical country like India.

Implementation Complexities
India's Social and Class Structures

India's bureaucracy, distinct from those in countries like the United States, Australia, New Zealand, or Canada, is deeply entrenched in class-based power dynamics, further intensified by Aadhaar. India has had its kings, contributing to a cultural perception of citizens as subjects to authority. The nation's diversity resembles a collection of multiple nations within one border. Ishaan described India as "a different planet to Australia," with its twenty-eight states and eight union territories, each subdivided into districts and smaller administrative divisions.

Each region has its own levels of government, economies, and policies regarding financial distribution to citizens. The Aadhaar system, with its biometric data storage, supposedly aims to ensure that

---

[127] So concerned was he (Ishaan) about government repercussions for his dissenting voice, he even avoided communicating with me on so-called private messaging on social media platforms, for example.

financial aid reaches the intended, highly nomadic recipients across this complex structure.[128]

In a relatively young democracy, classism persists, with power and wealth commanding undue influence. This dynamic is evident in various contexts, such as education, where the power to admit or deny student entry can create a master-servant relationship with the child's parents. The rule of law is often ineffective, leaving individuals without recourse and fostering a culture where everything has a price and power dictates outcomes. Sound familiar?

Bureaucratic inefficiency is prevalent, with government officials often avoiding work. If Aadhaar biometric systems fail, there is no accountability for the inconvenience caused to users. Ishaan, with equal measures of defeat and frustration, highlighted a government mindset that presumes citizens are guilty until proven innocent. This creates an imbalanced relationship where citizens have little power or recourse due to the unpredictable and selective application of the rule of law.

Conflict of Interest

This 'might-makes-right' dynamic leads to questionable associations. Take the case of Infosys co-founder, Nandan Nilekani, who took leave from the major technology company, and was appointed Chairman of the Unique Identification Authority of India (UIDAI) by the Indian government. Nilekani played a pivotal role in establishing Aadhaar, yet subsequent moves by Infosys draw into question the potential conflicts inherent in the overlap of private enterprise and public authority. (76)

For example, without an Aadhaar number in India, people encounter difficulties opening and accessing bank accounts. Despite the Supreme Court delaying Aadhaar linking deadlines, banks continue

---

[128] Census data from 2011 suggested 139 million Indians migrate domestically every year. (76)

to demand Aadhaar numbers and have been disabling accounts for non-compliance. Dr Anupam Saraph, an expert in governance of complex systems, blames the Finacle banking software from Infosys for mandating Aadhaar numbers in financial activities.

This highlights potential conflicts of interest and the negative impact on customer access to financial services. (77) The professor and author highlighted this conflict of interest in 2018, asking:

Infosys has a conflict of interest in pushing #Aadhaar, doesn't it? Is it a surprise that all new software from them seems to leave their customers (banks, government departments) helpless without forcing us, their customers, to submit Aadhaar? (78)

Faulty Assumptions

The benefits that supposedly outweigh these costs rest precariously on a number of assumptions. First of these is that the system's enforcers are incorruptible and act altruistically. However, this is a fragile premise, as it overlooks the possibility of these entities exploiting the system for personal gain. Simply implementing Aadhaar does not solve the underlying issue of bureaucratic corruption; it could just change how corruption appears.

The belief that Aadhaar can eliminate corruption fails to consider that bad behaviour might just migrate to other parts of the system. Digital ID can create new weaknesses and opportunities for fraud, especially for those skilled in manipulating digital environments. Digitising our identity does not remove the human factor, frequently the most vulnerable point in the fight against corruption.

The argument against corruption also overlooks its persistence in digital realms. Coercive forces in the physical world can compel individuals to make legitimate payments to their oppressors, and Aadhaar cannot prevent this. Additionally, scams exploiting sensitive information, like Aadhaar-linked bank details, persist and show that digital systems remain vulnerable to social engineering by fraudsters.

Aadhaar's rollout has also introduced economic and social friction, potentially outweighing the very problems we're told it seeks to address. Mandatory Aadhaar linkage has resulted in service exclusions and denials due to authentication issues, disproportionately impacting vulnerable groups and raising significant social costs.

Finally, the significant expenses involved in setting up and operating the Aadhaar system, including its creation, management, and security, as well as addressing errors and breaches, call into question the efficiency and cost-saving justifications for its implementation. Regardless, Ishaan's take on Aadhaar's India is clear:

The ideation for systems like this, comes from this thought: stopping corruption. Yet corruption in India has only worsened since Rajiv Gandhi's statement. We can't know whether the original motivations were genuine or not.

Fraud as a Service

How can we assess Ishaan's critique that Aadhaar's intended benefits contrast sharply with the reality of persistent fraud in India? A start is the city of Jamtara in India's East. There, the economy thrives on socially accepted scamming, known locally as the 'SIM card business'. Jamtara is one of many cities where fraudulent activity, supposedly mutually exclusive with centralised government digital ID, continues to provide ill-gotten income for many, particularly unemployed youth.

Ishaan recounted his 58-year-old illiterate mother who receives weekly scam calls from individuals posing as bank customer support. These callers fabricate issues requiring her credit card or passcode details to resolve, threatening to lock her card or account if she does not comply. This scam is widespread: in April 2023, six individuals from Jamtara were arrested for impersonating customer care representatives from reputable banks, defrauding 2,500 people of nearly USD 120,000 using 12,500 pre-activated SIM cards. (79)

With a nation of millions who are unaware of the most basic digital self-protection, Ishaan's conclusion was:

Digital illiteracy combined with corruption is a bad mix.

He pointed out that comparing Aadhaar to systems in countries with established, distributed governance is like comparing apples to oranges. In India, where there's often chaos due to the lack of such systems, Aadhaar could be seen as filling a gap. Conversely, in countries with a somewhat functional rule of law, and operational state-level systems, like those for issuing driver's licenses, the need for an Aadhaar-like approach is not clear. He asserted:

If a country already has effective processes in place,
why on earth would you want to go with an Aadhaar approach?

Identity Centralisation: A Hacker's Paradise

Who serves to commercially benefit from Aadhaar? Beyond the small-scale intimacy of credit card scam calls, Aadhaar's centralised nature raises profound concerns about enterprise level data security. Frequent reports of data breaches and unauthorised sharing of personal information for commercial gain, serve only to add to the growing apprehension among the people. One of India's most damaging data breaches occurred in October 2023, exposing a staggering 815 million Indians' personal details, including names, phone numbers, and Aadhaar numbers. (80) This centralised wealth of personally-identifying information was available for sale on the web. It took a US-based firm, Resecurity, to disclose the hack.

Private Collaboration

Aadhaar's design is not unique. Similar to other centralised systems, it outsources authentication and service delivery to third parties, akin to the 'companion entity' concept from the Australia Card proposal. This approach exposes Aadhaar's underlying identity data, increasing security risks and highlighting the crucial difference between raw data and its analysis, discussed in Chapter Four. While leaks of determinations are concerning, leaks of raw data are far more severe due to the alarming potential for reinterpretation and widespread misuse.

The inconsistent practices of external businesses have led to other significant Aadhaar data breaches, affecting a staggering 300 million, 20 million, and 100 million users in 2019, 2020, and 2021, respectively. (81) Eyes can easily glaze over when considering such vast numbers. Yet every one of those tens of millions of identity violations relates to a real person who must now face the consequences of forced participation in government digital ID. Those consequences can be significantly profound and personal.

The Human Cost

What is the underlying human cost? Aadhaar highlights the debate on inclusion and access in digital ID systems. Proponents argue that digital ID can enfranchise marginalised people lacking traditional identification by granting them access to essential services. While digital ID may indeed liberate people without traditional identification, this benefit is directly counterbalanced by the emerging risk of excluding people without access to technology or lacking sufficient digital literacy. Furthermore, digital inclusion could be realised through decentralised, market-driven private sector solutions, which, due to competition, are incentivised to develop accessible and user-friendly systems, even for the illiterate. Could this potentially make government involvement superfluous?

Two Phones, Two Lives

While discussing the effects of Aadhaar on his countrymen, Ishaan held up two phones, including an AUD $100 Nokia housing his Indian SIM card. He strongly emphasised the critical importance of maintaining an Aadhaar-linked phone number, even for expatriates, highlighting the vulnerability of SIM cards to damage or loss.

Losing the SIM card linked to Aadhaar can have far-reaching consequences, effectively erasing one's identity and corresponding access to services. The process of updating a phone number in the Aadhaar system is unclear and quite daunting, leading Ishaan to express a deep concern for the potential risks, stating:

> *I fear what can go wrong. I shouldn't need to have this fear as a citizen.*

The Challenge of Accessing Your Own Information

Ishaan re-counted a simple example from his life when trying to withdraw his superannuation before moving to Australia. Systemic failures often thwart this process with many employees struggling to access their Employees' Provident Fund (EPF) after job changes or post-employment. Many face frustration due to poor support from past employers and communication barriers with the Employees' Provident Fund Organisation (EPFO).

Challenges in managing EPF accounts can arise from discrepancies between EPF details—like name, father's name, date of birth, and Aadhaar data—and those on ID documents. Problems also occur due to factors outside employee control, such as employers not depositing contributions, or mismatches in recorded and actual employment dates. (82)

Leveraging his technical education, Ishaan meticulously completed all requirements for Aadhaar, crucial for superannuation withdrawal. Fully aware that proper engagement with the system was key to avoid delays, he ensured his mobile number and spelling matched government records where he was able to determine what these were. With equal measure of luck and attention to detail, he was able to withdraw his superannuation without the need for multiple 300-mile trips to a government office—sadly, many are not as fastidious or fortunate.

I Enjoyed Watching You Buy that Bra

To maintain balance in my exploration, I consulted with a technology entrepreneur and business owner currently working in the digital ID sector in India. For anonymity, we'll refer to him as Kabir. Kabir shared deep insights into the stress Aadhaar has inflicted on his cousin, highlighting the personal impacts of the system.

Because one often can't even purchase a bra in a shop without revealing your Aadhaar-linked phone number, she has faced harassment from stalkers who obtained her contact details through retail transactions. This could involve shop employees, nearby customers, or anyone who had access to her number once it entered the digital system. A hidden fallout of centralised digital ID systems then, is that of women being harassed by stalkers who call to say:

I enjoyed watching you buy that bra.

Emboldened by access to their personal information after mundane activities like shopping, policymakers often overlook this stark reality. It underscores the highly invasive nature of these systems that leave victims feeling profoundly violated.

Compounding the issue, residential addresses are not safe from the grasp of these systems, creating situations akin to the 1996 horror movie, *Scream*, where stalkers could make calls from outside the victim's bedroom window. People living under such systems face these and other repercussions of compulsory state involvement in identity management on a daily basis. Despite being a critical issue of our era, it does not receive the attention and consideration it warrants. Yet in extreme circumstances, addresses can be changed. Biometrics, on the other hand, cannot.

Aadhaar's Biometric Bind

Once compromised, biometric data like retinas, faces, and fingerprints cannot be easily changed and can be exploited maliciously, affecting people wherever they go. Despite these risks, the use of biometric identification with Aadhaar has surged without apparent caution.

Commercially available products, such as AtomX's facial recognition ticketing system, use Aadhaar biometrics to grant access to events, like stadium concerts, by verifying attendees' immutable features. AtomX champions their ability to "get people in quicker." One of their access control portals is capable of processing 2,500 entries per hour, a rate that is only ten percent of that possible with *Dehomag* sorting machines. (83) Although the system efficiently identifies ticket holders by their faces, the significant privacy risks it introduces could potentially lead to data misuse for exclusion as well as admission.

Access management is not a trivial undertaking and making granular, centralised decisions in real-time based on biometrics can fundamentally alter societal trust and freedom. Being denied legitimate access to a concert is one thing, to food is another. Yet to Aadhaar, they are potentially one and the same. Living without Aadhaar is almost impossible, leading to abuses including denial of employment, welfare, and food rations for those who experience technical issues with it. These exclusions have severe consequences, including starvation among single mothers in this modern, nuclear armed nation. (84)

Starvation: A Potential Outcome from a Failed System

Should the sight of people starving prompt immediate suspension and scrutiny of the responsible systems? These issues are not merely theoretical. They pose real hardship for mothers like Sita from Karkala village in India's Northwest. Despite her efforts to comply with Aadhaar, her fingerprints simply were not accepted by the digital arbiter, denying her work and her children rations. As a single woman without alternative income, she faced hunger. Her frail situation highlights the failure of a safety critical system, yet official responses lack the necessary urgency.

Air-crash investigators scrutinise smouldering wreckage to better understand failures in aviation safety systems. However, the tragedy of an air crash affecting hundreds often seems to resonate more deeply than the suffering of many more due to flawed government digital ID systems.

Is this purely a government issue? No! The webbed entanglement of private companies in the Aadhaar ecosystem serves to add further complexity, creating a maze of authentication and business processes stacked on top of one another, that leave many feeling ensnared in an inescapable labyrinth. Compulsory linking of Aadhaar to essential services, without alternatives, monopolises digital ID, forcing individuals to compromise their data and privacy to a centralised authority. Furthermore, individuals who attempt to voice concerns over the issue are faced with other risks.

Punishment for Dissent

The Cost of Criticism in Aadhaar's India

The consequences of dissent extend far beyond inconvenience, with intellectuals and activists facing silencing or imprisonment for criticising the system. In India, there is an unspoken threshold of acceptable public criticism, which can be easily misjudged. Instances of academics being barred from entering the country due to revoked visas are not rare. Take Professor Ashok Swain, based in Sweden, whose Overseas Citizenship of India (OCI) card was revoked by the government in February 2022, apparently due to his critical academic work on government policies. Swain astutely characterised the government's order as:

> ... *passed without any application of Judicial Mind, to the extent that it deems to be a routine/mechanical exercise of power. (85)*

Justice Prasad of the Delhi High Court supported Swain's notion, directing the government in July 2023 to substantiate its actions. (85)

Targeting Aadhaar's critics is not uncommon. Kabir bluntly told me that if he walked around in public with an "I hate Aadhaar" sign, he might not be walking around, freely at least, much longer. In another example, I learned about researchers from a university in the country's North who were involved in criticising Aadhaar through a philosophical lens. Their fate for questioning the system? They are all now serving jail sentences. While state governments enforce these actions, they appear to align with the central government's support of the national digital ID system, indicating the high stakes involved in Aadhaar's imposition on the population.

Kabir observed that he had never seen college professors being jailed, even going back five decades, highlighting the government's intent to rapidly suppress even reasoned critiques of Aadhaar. Incarceration remains a timeless tool for tyrants, underscoring the system's interest in self-preservation by stifling free expression and fostering a climate of fear and intimidation.

The Unyielding Grip of Centralised Identity Systems

What does India's future entail? According to Kabir, without major political changes, Aadhaar's integration may become permanent.

India's unique cultural context has allowed Aadhaar to become deeply rooted, especially in the business sector. Typically, only the government has the power and authority to undo such entrenched systems. Therefore, it appears the Indian people seem set to follow this path, however flawed, for the long term. They have my deepest sympathies.

Giving the Devil his Due

Despite these concerns, it is important to acknowledge that within India's intricate landscape, Aadhaar has the potential to drive digital economy growth, spur innovation, and simplify online services. Given the prevalence of fraud, these IDs could reap some efficiencies, facilitating smoother interactions between the government and the people.

Digital ID systems such as Aadhaar are often praised for making transactions more efficient and boosting the economy. Yet, their effect on government transparency is understandably complex. While they could improve accountability and decrease corruption through audit trails, the need for centralised systems to self-regulate may result in inconsistent enforcement, compromising intended transparency. Additionally, even if Aadhaar reduces corruption in some areas, the outcome may be that corruption is simply relocated elsewhere within the economy, complicating the evaluation of its overall impact on transparency.

The Indian government invested heavily to extend internet access to remote areas for Aadhaar, even distributing free mobile phones and SIM cards through the Reliance group. This significant investment could suggest hidden advantages for the government, likely related to increased centralised control and power, or a genuine benefit for otherwise disconnected communities. Which is correct? We will never be sure.

Regardless, surveillance capitalism can emerge as a derivative economy of such efforts, where personal data becomes a commodity for profit, often without consent. This serves to distort economic incentives and stifle true innovation and growth. The aim should be to adopt technologies that support genuine transactions, without enabling the misuse of personal data for peripheral economic gains.

Learning from India

Does merging identity management with control, surveillance, and policing endanger the foundation of freedom and draw into question the true beneficiaries of such systems? In countries with weak accountability and inefficient bureaucracies, centralising identity management may seem advantageous. Conversely, countries with robust, decentralised governance and less corruption should approach centralisation with caution. While each system is unique, it is arguably fair to consider giving India's approach some leeway, even though it faces challenges common to all centralised identity systems. Other countries considering central digital ID systems, however, must heed India's experience.

Who Drives Modern Agendas?

Representative democracies should reflect community-sourced ideas and values through legislative action. The Australia Card's resounding rejection supports this notion. However, the trajectory of centralised identity systems often shows a reversal of this process, with governments attempting top-down identity control, forcing citizens to respond defensively with highly unpredictable results.

Who keeps promoting these systems? This is not symptomatic of a governance structure providing representation of public will. For example, the revulsion with which the community rejected the Australia Card, firmly suggests the idea was not sourced there.

The deception exposed by the Joint Select Committee supports this idea.

This scenario suggests a governance failure, with elite-driven ideas imposed on the public without their knowledge. The lack of wide-spread public discourse on digital ID hints that it is certainly not a priority for most of the population. If the revelation that citizens' interests are not being adequately represented requires a Joint Select Committee, then the system simply cannot be functioning as it should within a truly representative democracy.

If true, this situation demands continuous, vigilant public effort to monitor and counteract harmful legislation with hidden agendas. We must scrutinise the origins of oppressive policies, and question why the public bears the burden of opposing them. It is highly problematic when ideas are top-down impositions, rather than bottom-up grassroots reflections of societal values within the legislative process, truly embodying public sentiment.

How can we account for the global similarities in these systems? The striking resemblance among centralised identity registries like the *Volkskartei*, Australia Card, and Aadhaar, despite their different origins and cultural contexts, suggests they were not shaped by community needs. All were driven by incumbent governments and enforced through de-humanising principles and policing. Each instance also involved a variant of the companion system to ensure private sector compliance, penalised dissent, and selectively denied resources based on simplistic labels and justifications. In the context of this political activity, the preservation of individual anonymity is as critical now as ever, to protect private property and the very fabric of society.

It is also important to acknowledge that government digital ID is often justified by the need to meet international standards and enhance cross-border cooperation. This idea may help to partly explain the top-down adoption of these systems. Yet, while acknowledging that international cooperation is important, it can be achieved in numerous ways and should not come at the expense of individual privacy rights.

Decentralised, privacy-preserving technologies can also meet these standards, and avoid establishing a uniform system, which can be susceptible to global security threats. This suggests that, while we strive for international collaboration, we must ensure that the methods employed do not undermine the autonomy and privacy of individuals.

Yet as before, we are again forgetting the enduring lessons that history has taught us. Defeating the Australia Card only bought time, and this lapse has opened the door to a new, more potent digital threat. As the children of this generation swim in pools and ride bicycles during the summer, the exact same evil is again brewing. Armed with the power of modern digital technology, it is as fearsome as ever. It falls on us to (again) guarantee freedom for the next generations, so that they may do the same.

The Australian Digital ID Bill 2024

A Rushed Inquiry

The Australian Digital ID Bill 2024 was passed into law on May 16, 2024, despite public protests. First introduced to the Senate on November 30, 2023, the bill was quickly referred to the Senate Economics Legislation Committee for review. By February 2024, the committee recommended the bill's passage. Despite a limited, poorly-communicated, one-month public consultation period over the Christmas holidays, the inquiry received hundreds of submissions highlighting the concerns discussed in this book, with the inquiry acknowledging that:

Most individuals who made submissions to the inquiry opposed the bills either fully or in part. (1 p. 39)

Biometrics

The bill outlines broad data collection powers, including the use, and sharing of biometric data—a dangerous method of identification that the University of New South Wales criticised as "largely unproven and untested." (1 p. 51) This was soundly supported by the New South Wales Council for Civil Liberties, warning that inaccuracies in facial recognition technology could lead to the "denial of essential government services" just as Aadhaar had done to Sita in India. (1 pp. 51-52)

Voluntarism

The bill states that "creating and using a digital ID is voluntary." (86 p. 92) However, it also allows for exemptions where participating parties can require individuals to use a digital ID as a condition for providing services. This could potentially lead to scenarios where people are forced into the system, undermining its voluntary nature. Senators Canavan and Rennick stated that:

... there is no limitation on the Regulator's powers but the vague and open-ended requirement for a mandatory requirement to be 'appropriate to do so'. (1 p. 75)

Supporting this, it was further observed that under Clause 84, accredited entities are immune from civil and criminal liability in some cases when "not providing the accredited service." (1 p. 34) Yet perhaps more disturbing than the bill itself was some of the conduct of participants in the process. For example, it was reported that banks sought to:

... modify the proposed digital ID legislation to circumvent the explicit prohibition on the use of racial markers. (87)

Law Enforcement Use

Many people expressed concerns that law enforcement exemptions to access data from the digital ID system were too permissive. They emphasised that people should be able to use the system without the fear of unauthorised access by law enforcement or private entities that Ishaan lives with under Aadhaar today. Digital Rights Watch also strongly opposed re-purposing digital ID data to monitor people, arguing that such practices could lead to mass surveillance, eroding already flimsy public trust. (1 p. 53)

The Ongoing Global Issue

Is digital ID encroachment only unfolding in Australia? Certainly not! The digital arms race we recognised in Chapter Four continues to play out across the global context. In early 2024, a KLM flight from Montreal to Schiphol Airport marked a Canada-Netherlands pilot project requested by the European Commission, to streamline border checks with the so-called Digital Travel Credential. This digital document, derived from passport chip data, is being justified by aims to reduce queues and simplify border management. (88)

Simultaneously, the Estonian Ministry of the Interior has chosen to keep identity cards mandatory for all citizens until the EU digital wallet is introduced, despite suggestions for making them voluntary. This decision comes amidst discussions on a draft law aiming to offer more flexibility. (89)

Tomorrow's Future is Not Determined

A Call to Action

Can the citizen win? Yes! Laws can be changed, but there is much to do. Experience is often said to be the best teacher. Yet as humans, we have the peculiar habit of overlooking history's lessons, choosing instead to enjoy successes, and suffer failures firsthand. Regardless, understanding our past remains crucial to retain firm contact with our foundational values. Through education, we empower ourselves to draw connections between historical events and current trends.

Douglas Murray highlights this in his book, *The War on the West,* where he cites a 2020 survey that reveals a concerning modern trend among Americans. (90 pp. 80-81) Nearly half of those surveyed between the ages of twenty and forty were unable to name one camp or ghetto from the Second World War, while more alarmingly, twelve percent were simply unaware of the Holocaust.

Wrapping Up

This book is my attempt to bridge that knowledge gap, and the startling lack of awareness revealed by such surveys underscores the urgency of my mission. Yet recounting data about the Holocaust is insufficient. We must be capable of analysing that data to uncover its origins and identify when they reemerge. A deeper scepticism of the world being built around us frees us from needing to be told what to think and do. To that end, these pages also invite proliferating critical thought.

Having shared my insights, the next step is yours to take.

I have laid the groundwork, and now it is time to turn that knowledge into action, fostering a culture of vigilance and education. Remember my *Oma* and use my family's story to reveal the identity predator to those around you. Teach them why we must always keep an eye out for its reemergence. Motivate others to call it out for what it is when it reappears, from opaque corridors of power. Ask those you love to picture their children with a childhood like my *Oma's* ... and how terrifying that would have been.

I understand this may feel difficult. These are challenging issues, and it's tempting to downplay their risks. Yet, I urge you to persevere. We *must* prevent the government from gaining a foothold on the beachhead of our identities. The notion of universal identity registration and pervasive surveillance, as a panacea for societal issues, is a deceptive false-choice, one that proponents of such systems constantly promote.

These systems are unmoored to any rigid restrictions preventing them from manifesting their full potential as authoritarian tools. It was not, and still is not, the purview of government to arbitrate relationships and trust within a community. It is for communities to embrace their identities and confront challenges as they emerge.

Lastly, I thank and congratulate you for finishing this book. Doing so has now left you at an important crossroads. These pages carry a fragile ember of freedom that has been passed to you. Will you cup your hands around it, nurturing it to live another day, or will you let it blow out? There is no middle ground on this absolute topic.

Failing to act now means choosing a side and risking everything. Time is short, but there is still a chance to peacefully reject government digital ID systems as anathema to humanity. I urge you to join me. With your help, *yes yours* dear reader, we can together shape a positive legacy that will benefit future generations.

United, we can ensure that when history recounts the most significant corporate and government assault on global human identity, it will affirm that the *citizen won!*

References

1. Digital ID Bill 2023 and the Digital ID (Transitional and Consequential Provisions) Bill 2023 Report. *Parliament of Australia.* [Online] 02 2024. [Cited: 01 03 2024.] https://www.aph.gov.au/Parliamentary_Business/Committees/Senate/Economics/DigitalIDBills2023/Report.

2. Bongiorno, Frank. *The Eighties - The Decade That Transformed Australia.* Collingwood, Australia : Black Inc, 2015.

3. Baudot, Marcel. *The Historical Encyclopedia of World War II.* New York : Greenwich House, 1984.

4. Roth, Götz Aly and Karl Heinz. *The Nazi Census - Identification and Control in the Third Reich.* Philadelphia : Temple University Press, 2004.

5. Jackson, Annie. *Operation Paperclip: The Secret Intelligence Program That Brought Nazi Scientists to America.* s.l. : Little, Brown US, 2015.

6. Bombing of Dresden. *Britannica.* [Online] 06 02 2024. [Cited: 21 03 2024.] https://www.britannica.com/event/bombing-of-Dresden.

7. Discover your unique story with AncestryDNA. *Ancestry.* [Online] 2024. [Cited: 09 04 2024.] https://www.ancestry.com.au/dna/.

8. Find out what your DNA says about you and your family. *23andMe.* [Online] 2024. [Cited: 09 04 2024.] https://www.23andme.com/en-int/.

9. The Jordan B. Peterson Podcast: 394. A Conversation About God | Dr John Lennox. *Spotify.* [Online] [Cited: 19 03 2024.] https://open.spotify.com/episode/69jU27VWLuk3x2fAKlk9Vz?si=7rh1U_M_SGuv1Ve8r7IlpQ&nd=1&dlsi=c4381998e74b4a91.

10. Roche, Darragh. Election 2024 Poll: How Voters Feel About Key Issues. *Newsweek Magazine.* [Online] 19 07 2023. [Cited: 19 03 2024.] https://www.newsweek.com/election-2024-poll-how-voters-feel-about-key-issues-1813658.

11. Electronic Frontiers Australia. Access Card / National ID Card. *Electronic Frontiers Australia.* [Online] 10 02 2008. [Cited: 17 03 2024.] https://www.efa.org.au/Issues/Privacy/accesscard.html.

12. Report of the Joint Select Committee on an Australia Card. *Parliament of Australia.* [Online] 05 1986. [Cited: 17 03 2024.] https://www.aph.gov.au/Parliamentary_Business/Committees/Senate/Significant_Reports/auscard/report/index.

13. United States Holocaust Memorial Museum. Nadine Schatz. *United States Holocaust Memorial Museum.* [Online] [Cited: 17 03 2024.] https://encyclopedia.ushmm.org/content/en/id-card/nadine-schatz.

14. Lillian Goldman Law Library. Statement by Edouard Daladier, Premier, to the Nation, September 3, 1939. *Yale Law School. The Avalon Project. Documents in Law, History and Diplomacy.* [Online] [Cited: 17 03 2024.] https://avalon.law.yale.edu/wwii/fr3.asp.

15. Richards, Byron J. *Fight For Your Health - Exposing the FDA's Betrayal of America.* Tucson, Arizona : Truth In Wellness, LLC, 2006.

16. Hilberg, Raul. Auschwitz and the 'Final Solution'. [book auth.] Yisrael Gutman and Michael Berenbaum. *Anatomy of the Auschwitz Death Camp.* Washington, D.C : Indiana University Press, 1994.

17. U.S. Department of Health & Human Services. Hydrogen Cyanide - Emergency Department/Hospital Management. *Chemical Hazards Emergency Medical Management.* [Online] [Cited: 18 03 2024.] https://chemm.hhs.gov/cyanide_hospital_mmg.htm.

18. Shetty, Jay. Beautiful Life - Dr.Zach Bush ON The Importance of Gut Health & Why Dehydration...- Jay Shetty 2023. *YouTube.* [Online] 01 08 2023. [Cited: 18 03 2024.] https://youtu.be/Wxs9DGwoL2s?si=w4Nm9k8UJYHYQmby&t=1098.

19. Bilyeu, Tom. Debunking Success: Alcohol, Laziness & Social Media Addiction Isn't Holding You Back | Alex Hormozi. *YouTube.* [Online] 05 09 2023. [Cited: 18 03 2024.] https://www.youtube.com/watch?v=Tigt75AcLLA&t=4914s.

20. NSDAP Rassenpolitisches Amt (Racial Political Office). *Die Frage der Behandlung der Bevölkerung der ehemaligen polnischen Gebiete nach rassenpolitischen Gesichtspunkten (The question of the treatment of the population of the former Polish territories according to racial political aspects).* s.l. : Bundesarchiv (German Federal Archives), 1939. BArch NS 2/56.

21. Windley, Phillip J. *Learning Digital Identity.* Sebastopol : O'Reilly, 2023.

22. GBG. The State of Digital Identity 2022. *GBG.* [Online] [Cited: 28 01 2023.] https://www.gbgplc.com/en/identity-verification/the-state-of-digital-identity-2022.

23. Sicherheitshauptamtes (Reich Security Main Office). *Jahreslagebericht 1938 (1938 Annual Management Report).* s.l. : Bundesarchiv (German Federal Archives), 1939. R 58/1094.

24. Knappenberger, Brian. *Turning Point: The Bomb and the Cold War. Episode 1: The Sun Came Up Tremendous.* Netflix, 2024.

25. Einsatzgruppen: An Overview. *United States Holocaust Memorial Museum (USHMM)*. [Online] [Cited: 18 03 2024.] https://encyclopedia.ushmm.org/content/en/article/einsatzgruppen.

26. Steyr-Daimler-Puch AG. *KZ-Gedenkstätte Gusen (Gusen concentration camp memorial)*. [Online] [Cited: 18 03 2024.] https://www.gusen-memorial.org/en/The-Concentration-Camp-Gusen/Forced-Labour/Steyr-Daimler-Puch-AG-.

27. Greene, Robert. *The 48 Laws of Power*. London : Profile Books, 2000.

28. Dawkins, Richard. *The God Delusion*. s.l. : BLACK SWAN, 2016.

29. Tamahori, Lee. *Die Another Day*. Metro-Goldwyn-Mayer, 2002.

30. System of triangles. *Auschwitz-Birkenau State Museum*. [Online] [Cited: 18 03 2024.] https://www.auschwitz.org/en/history/prisoner-classification/system-of-triangles/.

31. The head of the Race and Settlement Main Office Staff Management / Dept. Angebliche Äußerung des Führers über die rassische Wertigkeit der Ukrainer (Alleged Statement by the Leader about the racial value of Ukrainians). Berlin : Bundesarchiv (German Federal Archives), 1943. BArch NS 47/31.

32. Resnick, Paul, et al. The value of reputation on eBay: A controlled experiment. *Springer Link*. [Online] 06 2006. [Cited: 03 04 2024.] https://link.springer.com/article/10.1007/s10683-006-4309-2.

33. Molyneux, Stefan. *Practical Anarchy*. s.l. : CreateSpace Independent Publishing Platform, 2017.

34. Spielberg, Steven. *Jurassic Park*. Universal, 1993.

35. Black, Edwin. *IBM and the Holocaust*. Washington, DC : Dialog Press, 2012.

36. Greenleaf, Graham. The Australia Card: towards a national surveillance system. *Australasian Legal Information Institute*. [Online] [Cited: 18 03 2024.] http://www2.austlii.edu.au/itlaw/articles/GGozcard.html.

37. Der Reichsminister des Innern (The Reich Minister of the Interior). *Volkskartei (People's Index)*. s.l. : Bundesarchiv (German Federal Archives), 1939. R 70/155.

38. Ministeriums des Innern (Ministry of the Interior). *Supplementing the police registration register with a card index organised by birth cohort (Volkskartei)*. s.l. : Bundesarchiv (German Federal Archives), 1939. R 70-Polen/155.

39. *Ministerial Paper Issue A of the Reich and Prussian Ministry of the Interior.* The Reich Ministry of the Interior. A, Berlin : Bundesarchives R-70 Polen 00155, 1939.

40. Deutsche Hollerith Maschinen Gesellschaft m. b. H. (Dehomag). *The Hollerith punch card process and its application.* s.l. : Bundesarchiv (German Federal Archives). R 2/18540.

41. Locating the Victims. *United States Holocaust Memorial Museum.* [Online] [Cited: 17 04 2024.] https://encyclopedia.ushmm.org/content/en/article/locating-the-victims.

42. Deutsche Hollerith Maschinen Gesellschaft m. b. H. (Dehomag). *Letter to the Reich Insurance Office dated February 17, 1934.* s.l. : Bundesarchiv (German Federal Archives), 1934. R 2/18540.

43. Teague, Vanessa and Frengley, Ben. Submission to the Consultation on Digital ID. *Digital Identity.* [Online] 17 12 2020. [Cited: 21 03 2024.] https://www.digitalidentity.gov.au/sites/default/files/2021-01/consultation01-vanessa-teague.pdf.

44. Fridman, Lex. Tucker Carlson: Putin, Navalny, Trump, CIA, NSA, War, Politics & Freedom | Lex Fridman Podcast #414. *YouTube.* [Online] 28 02 2024. [Cited: 19 03 2024.] https://youtu.be/f_lRdkH_QoY?si=JvwF_QUdrJBOCL_Z&t=7396.

45. Dreyfus, Suelette. *A Secret Australia: revealed by the WikiLeaks Exposes.* Clayton, Victoria : Monash University Publishing, 2000.

46. Australian Bureau of Statistics. Census Household Form. s.l. : Commonwealth of Australia, 2021.

47. CDC Bought Phone Data To Monitor Americans' Compliance With Lockdowns, Contracts Show. *ZeroHedge.* [Online] 17 03 2023. [Cited: 27 05 2023.] https://www.zerohedge.com/political/cdc-bought-phone-data-monitor-americans-compliance-lockdowns-contracts-show .

48. Australian Securities and Investments Commission. Changes to the Commonwealth unclaimed money laws. *Australian Securities and Investments Commission.* [Online] [Cited: 19 03 2024.] https://asic.gov.au/regulatory-resources/financial-services/unclaimed-money/changes-to-the-commonwealth-unclaimed-money-laws/.

49. Bergen, Dr. Diego von. Telegram from Dr. Diego von Bergen, Ambassador of the German Reich to the Holy See. s.l. : Bundesarchiv (German Federal Archives), 1933. R 43-11_176.

50.   Friel, Mikhaila. Prince Harry says wearing a Nazi uniform to a costume party in 2005 was one of the biggest mistakes of his life. *Insider.* [Online] 09 12 2022. [Cited: 19 03 2024.] https://www.insider.com/prince-harry-says-wearing-nazi-costume-was-mistake-2022-12.

51.   Tarantino, Quentin. *Inglorious Bastards.* Universal Pictures, 2009.

52.   Digital Social ID. *Digital Social ID.* [Online] [Cited: 19 03 2024.] https://www.digitalsocial.id/.

53.   Frankfurt School Blockchain Center. YouTube. *Industry Insights (CAC23B) - Startup Pitch: Digitalsocial.ID (Simon Molitor).* [Online] 19 11 2023. [Cited: 19 03 2024.] https://www.youtube.com/watch?v=xDUtbbuD2CY.

54.   Infrastructure for Trusted Data markets. *Cheqd.* [Online] [Cited: 19 03 2024.] https://cheqd.io/.

55.   Trusted Digital Identity is Key for Your Next Big Idea. *Polygon ID.* [Online] [Cited: 19 03 2024.] https://www.polygonid.com/.

56.   The Ultimate Stack for Identity Management. *Hypersign.* [Online] [Cited: 19 03 2024.] https://www.hypersign.id/.

57.   Antadze, Lasha. Self-sovereign identity is not enough. *Blockworks.* [Online] 15 01 2024. [Cited: 19 03 2024.] https://blockworks.co/news/identity-eu-citizens-privacy-surveillance.

58.   Department of Homeland Security. Real ID. *Department of Homeland Security.* [Online] [Cited: 19 03 2024.] https://www.dhs.gov/real-id.

59.   Alliance for Responsible Citizenship. The most inspiring speech Jordan Peterson has ever given. *YouTube.* [Online] 02 11 2023. [Cited: 07 11 2023.] https://youtu.be/84kKxtZI0l0.

60.   Rothbard, Murray N. *Anatomy of the State.* s.l. : bnpublishing.com, 2014.

61.   Bormann, Martin. NSDAP Chief of Staff Internal Party Letter. Munich : Bundesarchiv (German Federal Archives), 1935. NS 6/219.

62.   President of the Reich Agency for Employment and Unemployment Insurance. Einführung des Arbeitsbuches (Introduction of the Workbook). Charlottenburg, Berlin : Bundesarchiv (German Federal Archives), 1935. NS 6/219.

63.   Chief of German Police. Berlin : Bundesarchiv (German Federal Archives), 1942. NS 47/31.

64.   NSDAP Race and Settlement Main Office. Treatment of Pregnant Foreign workers and Those Born in the Reich of Foreign Workers. s.l. : Bundesarchiv (German Federal Archives), 1944. NS 47/31.

65. Identity card (Deutsches Reich Kennkarte) issued to Margarete Sara Jacobsohn and stamped with a red letter J for "Jude" (Jew). *United States Holocaust Memorial Museum.* [Online] 17 05 2013. [Cited: 13 03 2023.] https://collections.ushmm.org/search/catalog/pa1131472.

66. Arnold Schwarzenegger says 'screw your freedom' to anti-maskers. *USA Today.* [Online] 12 08 2021. [Cited: 21 03 2024.] https://www.usatoday.com/videos/entertainment/celebrities/2021/08/12/screw-your-freedom-arnold-schwarzenegger-calls-out-anti-maskers/8111464002/.

67. Edward Westermann. Einsatzgruppen. *Britannica.* [Online] 28 02 2024. [Cited: 21 03 2024.] https://www.britannica.com/topic/Einsatzgruppen.

68. His Holiness John Paul II Short Biography. *The Holy See.* [Online] 30 06 2005. [Cited: 21 03 2024.] https://www.vatican.va/news_services/press/documentazione/documents/santopadre_biografie/giovanni_paolo_ii_biografia_breve_en.html.

69. Privacy International. On Campaigns of Opposition to ID Card Schemes. *Privacy International.* [Online] 01 1996. [Cited: 21 03 2024.] https://privacyinternational.org/sites/default/files/2017-12/ID%20CardSchemes.pdf.

70. Gallipoli landing. *National Museum Australia.* [Online] [Cited: 21 03 2024.] https://www.nma.gov.au/defining-moments/resources/gallipoli-landing.

71. John Simpson Kirkpatrick: Simpson and his donkey. *Australian War Memorial.* [Online] [Cited: 21 03 2024.] https://www.awm.gov.au/articles/encyclopedia/simpson.

72. Australia Card Bill 1986. *The Parliament of the Commonwealth of Australia House of Representatives.* [Online] [Cited: 03 05 2023.] https://parlinfo.aph.gov.au/parlInfo/download/legislation/billshistorical/HBILL198587V200009/upload_binary/009%20-%20AUSTRALIA%20CARD%20Bill%201986.pdf;fileType=application%2Fpdf.

73. Davies, Simon. The Loose Cannon: An overview of campaigns of opposition to National Identity Card proposals. *Australian Privacy Foundation.* [Online] [Cited: 28 1 2023.] https://www.privacy.org.au/About/Davies0402.html.

74. The Formation of the Australian Privacy Foundation. *Australian Privacy Foundation.* [Online] [Cited: 21 03 2024.] https://privacy.org.au/about/history/formation/.

75. 'Only 15 paise reaches the needy': SC quotes Rajiv Gandhi in its Aadhaar verdict. *Hindustan Times.* [Online] 11 06 2017. [Cited: 02 04 2024.] https://www.hindustantimes.com/india-news/only-15-paise-reaches-the-needy-sc-quotes-rajiv-gandhi-in-its-aadhaar-verdict/story-I8dniDGXF6ksulggTDgb9L.html.

76. Mahajan, Neelima. How Nandan Nilekani rolled out the authentication process that gave India its first standardized ID card. *Roland Berger.* [Online] 09 08 2018. [Cited: 02 04 2024.] https://www.rolandberger.com/en/Insights/Publications/The-making-of-India%E2%80%99s-biometric-Aadhaar-ID-program.html.

77. Aadhaar: Is Infosys founder Nilekani's conflict of interest making lives difficult for bank customers through Finacle? *Money Life.* [Online] 20 06 2018. [Cited: 02 04 2024.] https://www.moneylife.in/article/aadhaar-is-infosys-founder-nilekanis-conflict-of-interest-making-lives-difficult-for-bank-customers-through-finacle/54408.html.

78. Saraph, Dr. Anupam. AnupamSaraph. *Twitter.* [Online] 20 06 2018. [Cited: 02 04 2024.] https://twitter.com/AnupamSaraph/status/1009250773459394562.

79. How 2,500 people were duped by 6 men in Jamtara posing as customer care officials. *The Indian Express.* [Online] 20 04 2023. [Cited: 02 04 2024.] https://indianexpress.com/article/cities/delhi/jamtara-fraud-over-2500-persons-from-across-india-duped-8565860/.

80. Aadhaar data leak | Personal data of 81.5 crore Indians on sale on dark web: report. *The Economic Times.* [Online] 31 10 2023. [Cited: 21 03 2024.] https://economictimes.indiatimes.com/tech/technology/aadhar-data-leak-personal-data-of-81-5-crore-indians-on-sale-on-dark-web-report/articleshow/104856898.cms.

81. Hypersign. The solution to Aadhaar Data Leak Problem. [Online] 01 11 2023. [Cited: 21 03 2024.] https://www.hypersign.id/blogs/tpost/f1hosk1831-the-solution-to-aadhaar-data-leak-proble.

82. Unable to withdraw or transfer PF money? Here is how to lodge your grievances. *Financial Express.* [Online] 18 08 2020. [Cited: 02 04 2024.] https://www.financialexpress.com/money/unable-to-withdraw-or-transfer-pf-money-here-is-how-to-lodge-your-grievances/2058718/.

83. AtomX. Unlock The Power Of Cashless Payments. *AtomX.* [Online] [Cited: 21 03 2024.] https://atomx.in/.

84. Dey, Nikhil and Roy, Aruna. Excluded by Aadhaar. *The Indian Express.* [Online] 05 06 2017. [Cited: 21 03 2024.] https://indianexpress.com/article/opinion/columns/excluded-by-aadhaar-4689083/.

85. Delhi HC Sets Aside Order That Revoked OCI Status of Professor Ashok Swain. *The Wire*. [Online] 10 07 2023. [Cited: 02 04 2024.] https://thewire.in/law/delhi-hc-sets-aside-order-that-revoked-oci-status-of-professor-ashok-swain.

86. Digital ID Bill 2024 (Previous Citation Digital ID Bill 2023). *Parliament of Austrlia*. [Online] 30 11 2023. [Cited: 01 04 2024.] https://www.aph.gov.au/Parliamentary_Business/Bills_LEGislation/Bills_Search_Results/Result?bId=s1404.

87. Bajkowski, Julian. Digital ID bill passes senate. *The Mandarin*. [Online] 28 03 2024. [Cited: 02 04 2024.] https://www.themandarin.com.au/242920-digital-id-bill-passes-senate/.

88. Sing, Evie Kim. Passengers on KLM flight from Montreal to Amsterdam practise using DTC. *Identity Week*. [Online] 01 03 2024. [Cited: 02 04 2024.] https://identityweek.net/passengers-on-klm-flight-from-montreal-to-amsterdam-practise-using-dtc/.

89. Sing, Evie Kim. Estonia retains mandatory ID-cards amid mobile solutions debate. *Identity Week*. [Online] 04 03 2024. [Cited: 02 04 2024.] https://identityweek.net/estonia-retains-mandatory-id-cards-amid-mobile-solutions-debate/.

90. Murray, Douglas. *The War On The West*. London : Harper Collins, 2022.

91. Vronsky, Dr. Peter. Lecture 11 [Part 1] History of the Third Reich: The Final Solution Part 1. *YouTube*. [Online] 28 03 2020. [Cited: 15 03 2023.] https://www.youtube.com/watch?v=xNquJ3YKzvY.

92. Episode 1. Here We Go Again On Steroids. *Never Again Is Now Global*. [Online] [Cited: 15 03 2023.] https://neveragainisnowglobal.com/episode-1/.

93. Australian Government. About myGov. *myGov*. [Online] [Cited: 04 06 2023.] https://my.gov.au/en/about.

94. BMH Rinteln. *QARANC - Queen Alexandra's Royal Army Nursing Corps*. [Online] [Cited: 18 03 2024.] https://www.qaranc.co.uk/bmhrinteln.php.

95. Ramsey, Michael. Privacy infringement fears after police access data from SafeWA contact tracing app. *7News*. [Online] 15 06 2021. [Cited: 04 04 2024.] https://7news.com.au/news/western-australia-police/wa-police-accessed-contact-tracing-data-c-3118713.

96. Gallagher, Katy. Digital ID Bill Passes Senate. *Katy Gallagher*. [Online] 27 03 2024. [Cited: 05 04 2024.] https://www.katygallagher.com.au/media-centre/media-releases/digital-id-bill-passes-senate/.

97. Australian Bureau of Statistics. Australian Bureau of Statistics. *Regional population*. [Online] [Cited: 05 04 2024.] https://www.abs.gov.au/statistics/people/population/regional-population/latest-release.

98. Don't Kill Your Grandma. *Glossy.* [Online] 19 01 2021. [Cited: 05 04 2024.] http://glossyinc.com/2021/01/19/dont-kill-grandma/.

99. 'Don't kill granny' message for Preston youth aims to slow spread of Covid-19. *The Guardian.* [Online] 07 08 2020. [Cited: 05 04 2024.] https://www.theguardian.com/world/2020/aug/07/preston-added-to-areas-with-bans-on-households-mixing-due-to-covid-19.

100. Visentin, Lisa. National skills passport to provide digital ID for workers. *The Sydney Morning Herald.* [Online] 23 09 2023. [Cited: 05 04 2024.] https://www.smh.com.au/politics/federal/national-skills-passport-to-provide-digital-id-for-workers-20230921-p5e6i1.html.

101. ABC News. VicRoads apologises after thousands of Ballarat drivers sent emails with wrong surnames. *ABC News.* [Online] 29 06 2023. [Cited: 12 04 2024.] https://www.abc.net.au/news/2023-06-29/vicroads-digital-licence-email-blunder-ballarat/102540336.

102. Clark, Laine. Queensland digital driver's licence: Rollout is plagued by technical glitches and delays. *Daily Mail Australia.* [Online] 02 11 2023. [Cited: 12 04 2024.] https://www.dailymail.co.uk/news/article-12697411/Queensland-digital-drivers-licence.html.

103. Value of 1986 Australian Dollars today. *Inflation Tool.* [Online] [Cited: 17 04 2024.] https://www.inflationtool.com/australian-dollar/1986-to-present-value?amount=1000&year2=2023&frequency=yearly.

www.ingramcontent.com/pod-product-compliance
Lightning Source LLC
Chambersburg PA
CBHW062128020426
42335CB00013B/1136